SpringerBriefs in Cell Biology

For further volumes:
http://www.springer.com/series/10708

Ella Palmer

Cell-Based Microarrays

Review of Applications, Developments
and Technological Advances

 Springer

Ella Palmer
Faculty of Medicine, Quantitative Systems Biology
MRC Clinical Sciences Centre
Imperial College London
London, UK

ISBN 978-1-4939-0593-5   ISBN 978-1-4939-0594-2 (eBook)
DOI 10.1007/978-1-4939-0594-2
Springer New York Heidelberg Dordrecht London

Library of Congress Control Number: 2014934131

Printed on acid-free paper

Springer is part of Springer Science+Business Media (www.springer.com)

Preface

Cell-based microarrays are a technique originally described by the Sabatini group in 2001 as "reverse transfection." cDNA or siRNAs in vector constructs are printed onto a coated glass slide using a robotic microarrayer and the vector constructs are transfected in defined areas within cells grown over the surface of the slide or microplate. Downstream functional assays are then undertaken on the cells.

Cell-based microarrays have proven to be of vital importance, allowing high throughput analysis of over-expression and knock down of proteins. Since their development in 2001, they have advanced significantly to include over-expression arrays, short interfering RNA arrays, and compound and antibody arrays. The surface used to coat the glass slides has been researched and significantly improved to allow non-adherent cells to bind to the arrays. In an exciting new development, the cell-based array technology has been advanced by the use of microfluidic image cytometry for the analysis of small diagnostic samples with few cells.

This book succeeds the *Methods in Molecular Biology* series Protocols title *Cell-Based Microarrays* and is a much-needed review book. It contains chapters covering important aspects of the evolution of cell-based microarrays.

London, UK   Ella Palmer

Contents

Contributors

Priya Chudasama Division of Molecular and Experimental Surgery, University Medical Center Erlangen, Friedrich-Alexander University of Erlangen-Nuremberg, Erlangen, Germany

Robert Damoiseaux Molecular Screening Shared Resources, California Nanosystems Institute, University of California, Los Angeles, CA, USA

Dominika Gaus Division of Molecular and Experimental Surgery, University Medical Center Erlangen, Friedrich-Alexander University of Erlangen-Nuremberg, Erlangen, Germany

Andrew L. Hook Laboratory of Biophysics and Surface Analysis, University of Nottingham, Nottingham, UK

Ramona Jochmann Division of Molecular and Experimental Surgery, University Medical Center Erlangen, Friedrich-Alexander University of Erlangen-Nuremberg, Erlangen, Germany

Andreas Konrad Division of Molecular and Experimental Surgery, University Medical Center Erlangen, Friedrich-Alexander University of Erlangen-Nuremberg, Erlangen, Germany

Elisabeth Kuhn Division of Molecular and Experimental Surgery, University Medical Center Erlangen, Friedrich-Alexander University of Erlangen-Nuremberg, Erlangen, Germany

Elisabeth Naschberger Division of Molecular and Experimental Surgery, University Medical Center Erlangen, Friedrich-Alexander University of Erlangen-Nuremberg, Erlangen, Germany

Yoann Roupioz SPrAM, UMR 5819 CEA/CNRS/UJF-Grenoble 1, Institute for Nanosciences and Cryogeny, Grenoble, France

Michael Stürzl Division of Molecular and Experimental Surgery, University Medical Center Erlangen, Friedrich-Alexander University of Erlangen-Nuremberg, Erlangen, Germany

Westbrook M. Weaver Department of Bioengineering, University of California, Los Angeles, CA, USA

Frank Weise The Natural and Medical Sciences Institute (NMI), University of Tübingen, Reutlingen, Germany

Chapter 1
Cell-Based Microarrays: Recent Advances for Gene Function Analyses

Dominika Gaus, Elisabeth Kuhn, Andreas Konrad, Ramona Jochmann, Priya Chudasama, Elisabeth Naschberger, and Michael Stürzl

Abstract Reverse transfected cell microarrays (RTCM) are a powerful tool for the systematic analyses of gene functions. With this technology more than a thousand different nucleic acids can be transfected into eukaryotic cells in parallel on a single glass slide. This allows high-throughput analyses of gene functions using gain-of-function, loss-of-function, and mutation approaches. RTCM paved the way for genome-wide gene function analyses in order to determine gene functions involved in the molecular regulation of cell phenotypes in physiologic and pathophysiologic processes. Since RTCM was first introduced in 2001, the technique has been well established. The method was successfully used in several genome-wide and large-scale screenings, and novel analysis methods to detect gene functions have been developed. This chapter will summarize the most recent technological developments in the usage of RTCM, including optimization of (1) transfection efficiency, (2) reporter systems and automated data acquisition, (3) spotting density with decreased cross-contamination, and (4) the new development of assays to screen for paracrine gene effects.

D. Gaus • E. Kuhn • A. Konrad • R. Jochmann • P. Chudasama • E. Naschberger • M. Stürzl (✉)
Division of Molecular and Experimental Surgery, University Medical Center Erlangen,
Friedrich-Alexander University of Erlangen-Nuremberg, Schwabachanlage 10,
91054 Erlangen, Germany
e-mail: Dominika.gaus@uk-erlangen.de; Elisabeth.Kuhn@uk-erlangen.de;
Andreas.konrad@uk-erlangen.de; Ramona.jochmann@uk-erlangen.de;
Priya.chudasama@uk-erlangen.de; Elisabeth.naschberger@uk-erlangen.de;
Michael.stuerzl@uk-erlangen.de

E. Palmer, *Cell-Based Microarrays: Review of Applications,*
Developments and Technological Advances, SpringerBriefs in Cell Biology,
DOI 10.1007/978-1-4939-0594-2_1, © Ella Palmer 2014

Abbreviations

A549	Human alveolar basal epithelial cells
cDNA	Complementary DNA
CFP	Cyan fluorescent protein
COPI/COPII	Coat protein I/Coat protein II
COX-2	Cyclooxygenase 2
cPARP	Cleaved poly ADP ribose polymerase
CRE	cAMP response element
DNA	Deoxyribonucleic acid
dsRNA	Double-stranded ribonucleic acid
ECM	Extracellular matrix protein
EdU	5-ethynyl-2′-deoxyuridine
ER	Endoplasmic reticulum
GBP-1	Guanylate-binding protein 1
(E)GFP	(Enhanced) green fluorescent protein
GPCR	G protein-coupled receptor
GPR160	G protein-coupled receptor 160
HCT116	Colon carcinoma cell line
HEK293	Human embryonic kidney cell line
HEK239T	Human embryonic kidney cells expressing SV40 large T antigen
HeLa	Cervix carcinoma cell line
HFIB	Human fibroblasts
HIV	Human immunodeficiency virus
HT29	Colon adenocarcinoma cell line
HUVEC	Human umbilical vein endothelial cells
ICAM-1	Intercellular adhesion molecule 1
IFN-α	Interferon-α
IFN-γ	Interferon-γ
JNK	c-Jun N-terminal kinase
K562	Human erythroleukemia cell line
KPL-4	Breast cancer cell line
LICM	Lentivirus-infected cell microarray
LMP	Low melting point
MAPK	Mitogen-activated protein kinase
MCF-7	Breast cancer cell line
MicroSCALE	Microarrays of spatially confined adhesive lentiviral features
miRNA	Micro ribonucleic acid
NF-κB	Nuclear factor-kappa B
NLS	Nuclear localization signal
NPY	Neuropeptide Y
PC3	Prostate cancer cell line
PEST	Proline–glutamic acid–serine–threonine–protein sequence
PLL	Poly-L-lysine

RFP	Red fluorescent protein
RNAi	Ribonucleic acid interference
RTCM	Reverse transfected cell microarray
S2R+	Schneider S2 embryonic drosophila cell line
SHARPIN	SHANK-associated RH domain interactor
shRNA	Short hairpin ribonucleic acid
siRNA	Small interfering ribonucleic acid
SW480	Colon adenocarcinoma cell line
TNF-α	Tumor necrosis factor-α
TORC1	Target of rapamycin complex 1
tsO45G	Temperature-sensitive CFP-coupled viral membrane protein
TUNEL	Terminal deoxynucleotidyl transferase dUTP nick end label
U2OS	Human osteosarcoma cell line
VCAM-1	Vascular cell adhesion molecule 1
WiDr	Colon adenocarcinoma cell line
YFP	Yellow fluorescent protein

1.1 Introduction

Nowadays, genome-wide sequence analyses can be accomplished rapidly and within an affordable cost range by next-generation sequencing approaches, based on the highly parallel sequencing of DNA fragments [1]. However, nucleotide sequences provide only very limited insights into the functions of the respective genes. Gene functions are commonly studied within gain-of-function experiments by overexpression of genes or by loss-of-function experiments via silencing of genes with RNA interference (RNAi) technology [2].

Whole-genome gene function analyses require several thousands of transfections as well as time-consuming procedures. Applying standard technology, this is associated with high costs for plastic ware and reagents. The costs increased exponentially with not only single gene effects, but combinatorial effects with two or more genes, in different cell types and under different types of external stimulation, in the focus of the study. This demanded the development of new techniques allowing high-throughput gene function analyses within an affordable time and cost range.

1.1.1 Reverse Transfected Cell Microarray

Ziauddin and Sabatini introduced the method of gene function analysis at the microarray level [3]. They printed the DNA of respective gene expression plasmids together with gelatin on glass slides. Subsequently, the slides were dried, exposed to a transfection reagent, and overlaid with adherent mammalian cells. After 40 h of incubation, the cells were fixed and expression and/or function of the respective gene product was analyzed using immunofluorescence. This procedure was called reverse transfected cell microarray (RTCM), because, in contrast to conventional transfection approaches, the DNA was seeded first and afterwards the cells were added.

As an alternative to the method described above, the transfection reagent was mixed with DNA and gelatin, before the mixture was printed on the slide [3, 4]. The latter approach is nowadays most commonly used. The dried slides can be stored for up to 15 months [5]. In the first decade of introduction, the RTCM technology has been used successfully by researchers worldwide and has been adapted to several different screening goals as reviewed previously (for review, see: [2, 4]) and shortly summarized below.

The optimization of transfection efficiency for various cell types was achieved by the use of different slide coatings, for example fibronectin for the transfection of primary human mesenchymal stem cells or collagen IV for the transfection of pheochromocytoma cells [6, 7].

Overexpression of genes on the cell microarray level lead to the identification of pro-apoptotic genes by analyzing the yellow fluorescent protein (YFP)-contrasted phenotypic changes associated with the late stage of apoptosis, such as apoptotic body formation, DNA cleavage, and fragmentation of the nucleus [8]. Apoptosis-regulating genes were also analyzed with the RTCM technology using a terminal deoxynucleotide transferase dUTP nick end label (TUNEL) assay [9].

Gene effects on different signaling pathways were analyzed by transfection of reporter constructs containing an indicator gene such as green fluorescent protein (GFP) or red fluorescent protein (RFP) under the control of a promoter activated by the signaling pathway of interest. In this way, functional screens for genes activating NF-κB [10], MAPK, JNK [11], or the CRE pathway [12] were performed.

Moreover, gene-silencing approaches were accomplished using the RTCM. Silva and colleagues successfully applied short hairpin RNAs (shRNAs) on a chip to identify genes involved in cytokinesis and proteasomal degradation [13]. In order to introduce shRNA or complementary DNA (cDNA) into hard-to-transfect primary cells and nondividing cells, Bailey and colleagues developed a lentivirus-infected cell microarray (LICM), where pseudotyped lentiviruses encoding either shRNA or cDNA were printed on a slide [14].

RTCM was also combined with time-lapse microscopy and computational phenotyping as it was demonstrated by a genome-wide RNAi approach studying genes involved in the regulation of the cell cycle [15, 16].

Furthermore, RTCM was adapted to the use with non-adherent cells. Kato and colleagues immobilized human erythroleukemia cells (K562) to a glass slide with a biocompatible anchor for membranes [17]. The immobilized cells were reverse transfected with either GFP- or RFP-encoding plasmids or siRNA and showed no interspot cross-contamination.

Recently, different research groups introduced further technological improvements of the RTCM technology. Most relevant developments will be summarized here and include (1) the increase of transfection efficiencies in many different cell types, (2) the optimization of reporter systems and automated data acquisition, (3) the increase of spotting density with reduced cross-contamination, and (4) the development of assays to screen for paracrine gene effects (Table 1.1).

Table 1.1 Recent advances in RTCM

Goal	Approach	Cell type	Reporter system	Coating	Gene transfer	Ref.
Transfection efficiency						
Optimization of gene/ substrate uptake	Generation of a silica film composed of an upright nanosheet network, adsorbing DNA/transfection-mix	HEK293	GFP expression	Silica-film composed of upright nanosheet	LyoVec	[18]
Transgene expression in a broad range of adherent cell types	Rapid and parallel approach for the purification of lentiviruses, which are spotted onto the slide and are used for cell transduction	27 of 32 different cell types transduced	GFP expression and immunofluorescence staining	Hydrogel-coated glass slides	Transduction with lentiviruses	[19]
Reporter systems						
Identification of human host cell factors required for *Trypanosoma cruzi* infection	Human genome siRNA library targeting 21,127 human genes impairing entrance and/or proliferation of *T. cruzi*	U2OS	Automated quantification of cell nuclei and number of parasites per cell (cell and parasite nuclei labeled with DRAQ5 and differentiated by size)	Glass coverslips	Lipofectamine 2000	[20]
Identification of host genes involved in the early steps of HIV-1 infection without affecting cellular growth or viability	Human genome siRNA library targeting 21,127 human genes screened for HIV-1 infection with 15 imaging parameters	Hela-CD4+ LTR-GFP	Automated acquisition of 15 different imaging parameters (HIV-1 infection monitored by GFP expression via LTR-activation and syncytia formation)	Glass coverslips	Effectene	[21]
Identification of proteins regulating the early secretory pathway	Screen of 51,000 siRNAs for interference with ER-to-plasma membrane transport of fluorescently labeled cargo protein (tsO45G)	HeLa	Automated detection of immunostained intracellular and cell surface tsO45G; in addition COPI, COPII, and GM130 were detected as indicators for the early secretory pathway	Chambered cover glass slides	Lipofectamine 2000	[23]
Identification of novel regulators of the TORC1-signaling pathway	Screen of 13,618 dsRNAs to identify regulators of the TORC-1-signaling pathway in *Drosophila* cells	S2R+ cell expressing human S6 ribosomal protein (S6_S2R+)	Immunofluorescent staining of human pS6 (target of rnTOR-activated S6 kinase, involved in cell size, cell growth, cell survival, and glucose homeostasis)	Poly-lysine glass slides	Not given	[25]

(continued)

Table 1.1 (continued)

Goal	Approach	Cell type	Reporter system	Coating	Gene transfer	Ref.
Time-lapse approaches						
Construction of a reporter to monitor live dynamics of transcriptional activity	Newly generated fluorophore reporter (Venus-NLS-PEST) with high signal intensity, fast maturation and degradation, and signal restriction to the nucleus	HEK293T	Fluorophore reporter downstream of inducible promoter	Chambered cover glass slides	Lipofectamine 2000	[26]
Increase of spotting density and reduced cross-contamination						
Increase of spotting density, optimization of signal-to background ratio, and minimization of interspot contamination in different screening approaches	Use of hydrophobic microplates, spotting of ECM proteins in tranfection mixtures, nonenzymatic cell dissociation prior to seeding, restricted cell adhesion time, and careful washings in order to cause exclusive attachment of cells onto transfection spots. 92 different cell types were analyzed. Screenings with different siRNA and miRNA libraries for regulators of proliferation [27, 29–31], apoptosis [27, 31], and β1-integrin activation [32]	92 different cell lines; 85 were successfully seeded and cultured on the transfection spots	Immunofluorescent staining of proliferation markers EdU and Ki-67, [27, 29–31], apoptosis marker-cleaved PARP [27, 31], and activated β1-Integrin [32]	Untreated polystyrene microplates [27, 29–32] or chambered cover glasses [28]	siLentFect [27, 29–32] or Lipofectamine 2000 [28]	[27–32]
Paracrine cell interactions						
Evaluation of paracrine cell interactions	Establishment of selective transfection of only one of the two different cell types simultaneously spotted and restriction of the diffusion of paracrine effectors	HEK293T as effector cell; human fibroblasts, colorectal carcinoma cell lines (WiDR, HT29), and endothelial cells (HUVEC) as indicator cells	Immunofluorescent staining of paracrine induced expression of GBP-1, COX-2, VCAM-1, ICAM-1	Glass slides	Lipofectamine 2000	[33]

1.2 Recent Technological Improvements of the RTCM Technology

1.2.1 Optimization of Reverse Transfection Efficiencies

Glass slides and microplates were frequently coated with organic materials such as gelatin, collagen, or poly-L-lysine in order to improve transfection efficiency [2]. Recently, also inorganic substrates such as silicon and titanium films were used for reverse transfections. Newly developed wafers with surfaces composed of vertically oriented parallel silica sheets in the nanometer range, which increase the surface, were shown to result in high reverse transfection efficiencies in HEK293 cells as monitored by GFP expression [18]. Cell adherence and morphology were not impaired on the silica film. However, the authors did not show how different cell types may behave and whether parallel transfection of different genes on one film may be feasible. Microarray spotting of different plasmids on these surfaces has to be proven.

For hard-to-transfect cells, transduction with lentiviruses can be used. However, the use of LICM for cells which are difficult to transfect harbors the disadvantage that large libraries of purified lentiviruses are essential, which requires ultracentrifugation and cannot be carried out in a highly parallel mode [14]. Due to this limitation, Wood and colleagues established a simple procedure for lentivirus purification, which can be performed in parallel in a multiwell range. Lentiviruses are precipitated and purified by the addition of a polymer complex (polyelectrolyte complexation) and printed onto coated slides. The respective procedure was called microarrays of spatially confined adhesive lentiviral features (MicroSCALE) [19]. This method allowed stable and selectable transgene expression for up to 14 days. The authors showed that 27 different cell lines of 32 analyzed in total were suitable for MicroSCALE. Focusing on 618 lentivirally overexpressed kinases, the authors screened for modifiers of the sensitivity of melanoma cells to clinically relevant therapies [19].

1.2.2 Optimization of Reporter Systems and Automated Data Acquisition

In the last decade, loss-of-function analyses have emerged as a potent tool also for RTCM, which became specifically accessible with the availability of commercially available siRNA libraries against the whole genome.

Using whole-genome siRNA libraries, Genovesio and colleagues were able to identify human host genes involved in infections with *Trypanosoma cruzi* and human immunodeficiency virus-1 (HIV-1) [20, 21]. In both studies, a human genome siRNA library targeting more than 21,000 human genes was spotted, followed by seeding of U2OS cells (infection by *T. cruzi*) or HeLa-CD4+ LTR-GFP cells (infection by HIV-1). Infection with *T. cruzi* was quantitatively determined by automated fluorescence microscopy counting the nuclei of infected cells and the numbers of parasites on individual transfection spots. Transfection spots were labeled by co-transfection of a fluorescent oligonucleotide. Using this approach, 162 candidate genes which affected *T. cruzi* infection were identified. Of note, 15 genes were validated in secondary screening approaches using classic methodology [20].

The screen for genes involved in HIV-1 replication was carried out as described above. HeLa-CD4+ LTR-GFP cells were used which indicate HIV-1 infection via GFP expression driven by the LTR. In this work the authors increased the stringency of the screening by including a higher number of replicas (seven) and inclusion of a high number of different parameters in the automated data acquisition [21]. These parameters included, among others, cell number, cell distribution, syncytia formation, GFP reporter intensity and local signal background. Using this setup, they identified 56 cellular genes, of which 45 were novel genes involved in HIV-1 infection [21].

In another genome-wide siRNA screen, Simpson and colleagues expanded earlier approaches [22] in order to identify genes involved in endoplasmatic reticulum (ER)-to-plasma membrane transport. As an indicator, they used a cargo protein of the secretory pathway, namely the temperature-sensitive viral membrane protein (tsO45G), coupled to cyan fluorescent protein (CFP) [23]. Using automated microscopy, they quantified the level of the fluorescently labeled protein inside the cell and at the plasma membrane at the single cell level. They identified 554 siRNAs which repressed the amount of membrane associated tsO45G and consequently may be involved in the secretion pathway. In a second RNAi-based screening, they monitored phenotypic effects on the coating proteins I and II (COPI, COPII) and the Golgi matrix protein GM130 as indicators of the early secretory pathway. By this approach, 331 of 554 identified genes were classified to the early secretory pathway [23].

Moreover, large-scale siRNA screens were performed in *Drosophila melanogaster* cells [24]. Lindquist and colleagues attempted to identify regulators of the TORC1-S6K-signaling pathway in a genome-wide screening using 13,618 dsRNAs [25]. A *Drosopophila* cell line stably expressing human S6 ribosomal protein, which is phosphorylated by the kinase S6K was generated. The phosphorylation of hS6 was monitored with immunofluorescence using a specific antibody, whereas no antibodies were available for the respective *Drosophila* protein. The screening revealed 70 novel regulators of p-S6 or S6K. Among those were genes involved in the MAPK pathway, lipid metabolism, cell cycle, and the proteasome [25].

1.2.3   Time-Lapse Approaches

Most RTCM cell microarray screenings were combined with endpoint measurements. Only recently, Neumann and colleagues combined RTCM technology with time-lapse microscopy [15, 16]. On the basis of this technique, Rajan and colleagues aimed to improve the reporter molecule for time-lapse approaches [26]. They were specifically interested in measuring live dynamic changes of the transcriptional activity of different promoters. Optimal reporter molecules should (1) exhibit a high signal intensity, (2) appear rapidly upon activation of the promoter, (3) disappear after promoter activity is ceased, and (4) be restricted to a regularly shaped cell compartment which can be easily detected by automatic microscopy. To this goal, they constructed a cDNA encoding a recombinant Venus-NLS-PEST reporter protein. The Venus polypeptide matures 15 times faster and is 30 times brighter than enhanced yellow fluorescent protein (EYFP), the PEST domain regulates rapid degradation of the protein in case transcription has stopped, and the nuclear localization signal (NLS) restricts the signal to the cell nucleus. The function of this reporter construct could be successfully demonstrated in pilot time-lapse experiments using three test promoters and co-transfecting different activators with the RTCM method and HEK293T cells [26].

1.2.4 Increase in Spotting Density and Reduced Cross-Contamination

Protocols which restrict cell adhesion to the transfection spots have been established in order to increase the spotting density and to reduce cross-contamination between neighboring transfection areas. This was achieved by spotting extracellular matrix (ECM) proteins on untreated polystyrene-coated microplates where the cells were seeded [27]. Polystyrene microplates are more hydrophobic compared to glass slides, thus prohibiting cell attachment. Moreover, nonenzymatic cell dissociation methods prior to cell seeding, reduced cell adhesion times, and careful washing steps were found to be necessary for the cells to grow exclusively on the transfection spots. It was shown that spot-to-spot cross-contamination was basically absent using this approach [28].

Rantala and colleagues tested 92 cell lines by using this protocol and demonstrated that 85 cell lines were successfully seeded and cultured on the arrays. Five of these cell lines [prostate epithelial cells (RWPE-1), prostate cancer cells (LAPC-4, PC3, VCaP), osteosarcoma cells (U2OS)] were efficiently transfected under these conditions [27]. Using this methodological setup, the expression of 492 different G protein-coupled receptor (GPCR) molecules was inhibited with siRNA. Subsequently, the effects on cell growth and survival were analyzed by immunofluorescent staining of Ki-67 (proliferation marker) and cleaved poly ADP ribose polymerase (cPARP, apoptosis marker) [27]. Neuropeptide Y (NPY) and GPR160 siRNAs were found to be the strongest inhibitors of prostate cancer cell growth indicating that both genes are potent activators of cell growth [27].

This new protocol was used successfully in four other screens. (1) Fredlund and colleagues validated previously classified cell cycle genes in breast cancer as involved in either mitotic checkpoint or mitotic progression. To this end, they used an siRNA library, targeting 5,760 selected genes in breast cancer cells (KPL-4) and stained for the proliferation marker Ki-67 [29]. (2) Cepeda and colleagues identified novel F-box proteins involved in tumor cell proliferation in four different types of tumor cells by spotting an F-box siRNA library and measuring 5-ethynyl-2'-deoxyuridine (EdU)-incorporation [30]. In a second siRNA screen, targeting 6,135 genes including 53F-box genes, the authors validated the first screening results measuring Ki-67 fluorescence intensity. In both screens, FBXO28, an ubiquitin ligase, was identified as a regulator of tumor cell proliferation [30]. (3) In a further screen, an miRNA precursor library containing 319 chemically modified double-stranded RNA molecules was used to mimic endogenous mature human miRNAs [31]. In the colorectal carcinoma cell lines HT29, HCT116, and SW480, the effects on cell proliferation (EdU-incorporation) and apoptosis (immunofluorescent staining of cleaved PARP) were analyzed [31]. Firty-three miRNAs were identified as oncogene-like miRNAs and 93 miRNAs as tumor-suppression-like miRNAs. (4) Integrin-inactivating kinases and/or phosphatases were investigated in PC3 prostate cancer cells using a commercially available kinase-phosphatase siRNA library ($n = 897$) [32]. A total of 2.5 % of all siRNAs increased the active integrin expression as determined by immunofluorescent staining for active β-integrins. Of these, SHANK-associated RH domain interactor (SHARPIN) was identified as a novel integrin inhibitor, which keeps integrins in an inactive state and prevents recruitment of integrin activators [32].

1.2.5   Paracrine Cell Interactions

The RTCM technique is commonly used to study intra- or autocrine effects. However, in the past few years, paracrine mechanisms became increasingly relevant, as indicated by the numerous publications addressing the role of the microenvironment in human diseases.

In order to investigate paracrine interactions between different cell types, either transwell or conditioned medium approaches are required as yet, which are both costly and time-consuming. To allow a high-throughput screening for putative paracrine gene activities, the RTCM procedure was modified for this purpose (Fig. 1.1) [33]. Genes of interest were spotted as in the original procedure, however, subsequently, two different cell types with low and high permissiveness for transfection were seeded onto the slide: (1) an effector cell which is transfected with the genes of interest and (2) an indicator cell, which is not transfected under the conditions used, in order to detect specific paracrine effects exerted from the transfected effector cells (Fig. 1.2a, b). Spot-to-spot diffusion of the paracrine mediators was prevented by a matrix overlay [2 % low melting point (LMP) agarose, Fig. 1.2c], ultimately allowing 192 parallel tests for paracrine gene activations on a single chip. As a read-out, typically an immunofluorescent staining of the indicator protein was performed. Alternatively, activation/inhibition of signaling pathways in the indicator cell can be monitored by using cells expressing a reporter protein under the control of a

Fig. 1.1 Scheme of the *Parachip*. (**a**) Slide with transfection spots (indicated by *yellow arrow*). (**b**) Overlay with effector cells (*grey*) and indicator cells (*light brown*). Indicator cells are selectively transfected (selective reverse transfection, *green*). (**c**) Overlay of a matrix (*pink*), which locally restricts diffusion of paracrine mediators (*red*) released from transfected effector cells. (**d**) Activation of indicator cells (*dark brown*) by the paracrine mediator. The transfected gene may code directly for the secreted paracrine mediator or may modify/activate a cellular factor in the effector cell, which is then secreted and functions as a paracrine mediator (*white arrows*). [Detailed protocol published in Kuhn et al. (2012)]

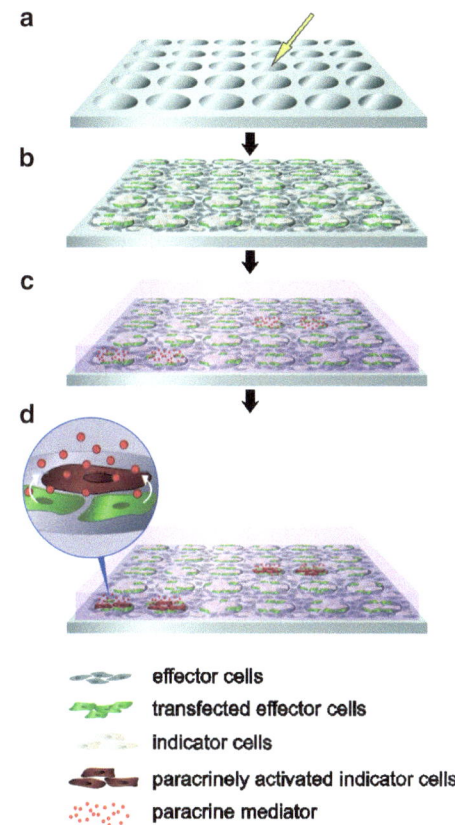

- effector cells
- transfected effector cells
- indicator cells
- paracrinely activated indicator cells
- paracrine mediator

promoter, which is activated/inhibited by the respective signal transduction pathway. The broad applicability and robustness of this technique was demonstrated using (1) HEK293T cells as indicator cells in combination with various responder cell types (fibroblasts, colorectal cancer cell lines and endothelial cells), (2) various paracrine inducers [tumor necrosis factor-α (TNF-α), interferon-α (IFN-α), and interferon-γ (IFN-γ)], and (3) various indicator genes [guanylate-binding protein 1 (GBP-1), intercellular adhesion molecule 1 (ICAM-1), vascular cell adhesion molecule 1 (VCAM-1), and cyclooxygenase 2 (COX-2)] [33]. This approach allowed for the first time a highly parallel analysis of paracrine gene functions and thus may facilitate the characterization of genes involved in heterotypic cell communication in a broad range of research areas. Of note, this technology allows identification of genes involved in paracrine interactions in multiple ways, including that the gene introduced into the effector cells (1) may code directly for a secreted product which activates the indicator cells or (2) may induce or modify in the effector cell another factor which subsequently is secreted and activates the indicator cells [33].

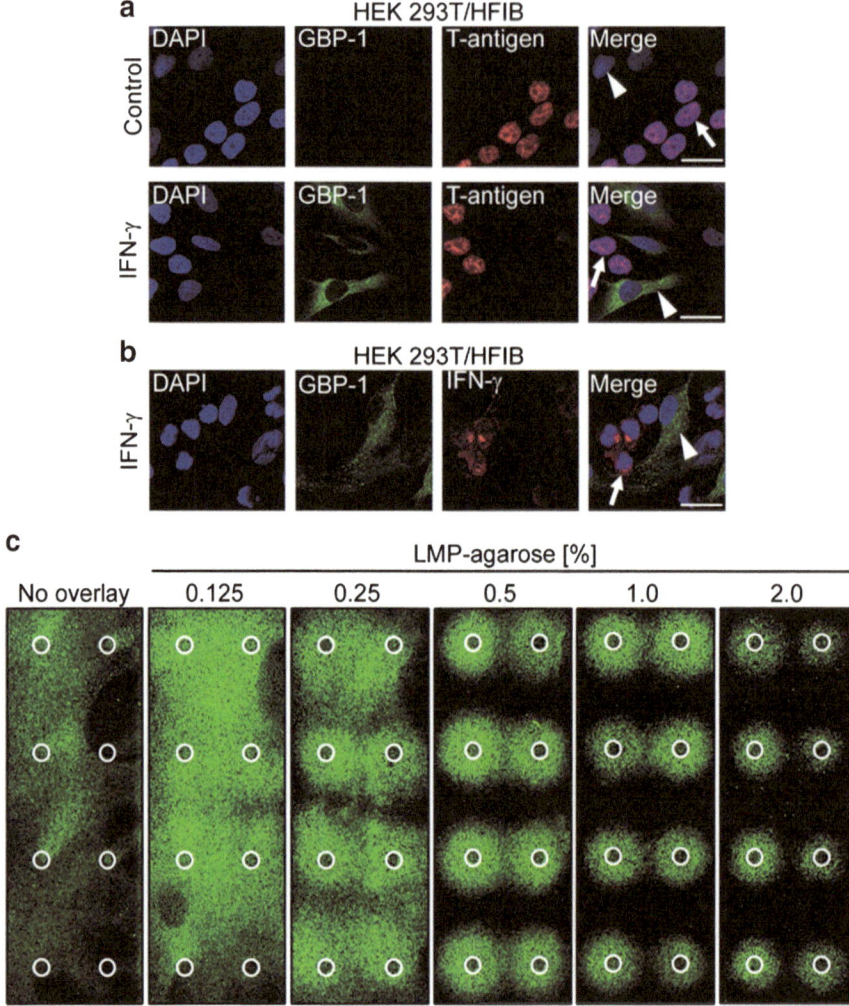

Fig. 1.2 Paracrine induction of the large GTPase GBP-1 by IFN-γ on a chip. An IFN-γ expression plasmid (IFN-γ) or empty vector (control) was spotted onto the chip. Subsequently, a mixture of HEK293T cells (effector cells, *arrow*) and human fibroblasts (indicator cells, *arrowhead*) were seeded. Selectively, HEK293T cells (**a**, T-antigen positive, *red*) expressed IFN-γ (**b**, *red*), whereas expression of the IFN-γ-induced GBP-1 was observed only in human fibroblasts (**a**, **b**, *green*). Panels show DAPI staining (*blue*) and immunofluorescence staining of GBP-1 (*green*), T-antigen (**a**, *red*), or IFN-γ (**b**, *red*). Scale bars=25 μm. (**c**) Identical approach as in (**a**) and (**b**), but the transfected cells were overlaid with low melting point (LMP) agarose in increasing concentrations to restrict the diffusion of IFN-γ secreted by HEK293T cells. Paracrine induction of GBP-1 (*green*) was restricted to the transfection spot (*indicated by circles*) when an LMP agarose overlay of 1.5 or 2 % was used. [Kuhn et al. (2012) A novel chip-based parallel transfection assay to evaluate paracrine cell interactions. Lab Chip 12: 1363–1372]. Reproduced by permission of The Royal Society of Chemistry (RSC). [Kuhn et al. (2012)]

1.3   Summary and Outlook

RTCM has been well established in the past years and has been successfully used for several gene function screens on a genome-wide level, both within gain-of-function and loss-of-function applications. Meanwhile, a high variety of different cell types can be subjected to this method and signal detection has been significantly optimized to allow automated data acquisition and processing. Innovative new screening methods for gene activities have been reported. Most approaches are presently focusing on autocrine gene activities, but in a first report, the application of RTCM in the analysis of paracrine gene functions within heterotypic cell–cell interaction, suggests that there may be perspectives for the development of novel screening methods analyzing multicellular interactions. RTCM is a miniaturization approach of cell biological methodology. This allows a highly parallel screening approach under identical conditions and at a clearly reduced cost basis as compared to classical approaches of cell biology. In the future, this may result in a similar "supernova" of new data in the field of gene function analyses as presently observed in the areas of gene sequencing and transcriptome analyses.

Acknowledgement We thank Mahimaidos Manoharan for valuable technical help and Alexander Jakin for help with the graphics.

Funding This work was supported by grants of the German Research Foundation (DFG-GK 1071, STU 238/6-1, SFB796), the German Federal Ministry of Education and Research (BMBF, Polyprobe-Study), the FAU Emerging Fields Initiative to M.S., of the German Cancer Aid (109510) to M.S. and E.N., and of the Interdisciplinary Center for Clinical Research (IZKF) of the University Hospital of the University of Erlangen-Nuremberg to M.S. and R.J.

References

1. Shendure J, Ji H (2008) Next-generation DNA sequencing. Nat Biotechnol 26:1135–1145
2. Stürzl M, Konrad A, Sander G, Wies E, Neipel F, Naschberger E, Reipschläger S, Gonin-Laurent N, Horch RE, Kneser U, Hohenberger W, Erfle H, Thurau M (2008) High throughput screening of gene functions in mammalian cells using reverse transfected cell arrays: review and protocol. Comb Chem High Throughput Screen 11:159–172
3. Ziauddin J, Sabatini DM (2001) Microarrays of cells expressing defined cDNAs. Nature 411:107–110
4. Konrad A, Jochmann R, Kuhn E, Naschberger E, Chudasama P, Stürzl M (2011) Reverse transfected cell microarrays in infectious disease research. Methods Mol Biol 706:107–118
5. Erfle H, Neumann B, Liebel U, Rogers P, Held M, Walter T, Ellenberg J, Pepperkok R (2007) Reverse transfection on cell arrays for high content screening microscopy. Nat Protoc 2:392–399
6. Yoshikawa T, Uchimura E, Kishi M, Funeriu DP, Miyake M, Miyake J (2004) Transfection microarray of human mesenchymal stem cells and on-chip siRNA gene knockdown. J Control Release 96:227–232

7. Uchimura E, Yamada S, Uebersax L, Yoshikawa T, Matsumoto K, Kishi M, Funeriu DP, Miyake M, Miyake J (2005) On-chip transfection of PC12 cells based on the rational understanding of the role of ECM molecules: efficient, non-viral transfection of PC12 cells using collagen IV. Neurosci Lett 378:40–43

8. Mannherz O, Mertens D, Hahn M, Lichter P (2006) Functional screening for proapoptotic genes by reverse transfection cell array technology. Genomics 87:665–672

9. Palmer EL, Miller AD, Freeman TC (2006) Identification and characterisation of human apoptosis inducing proteins using cell-based transfection microarrays and expression analysis. BMC Genomics 7:145

10. Konrad A, Wies E, Thurau M, Marquardt G, Naschberger E, Hentschel S, Jochmann R, Schulz TF, Erfle H, Brors B, Lausen B, Neipel F, Stürzl M (2009) A systems biology approach to identify the combination effects of human herpesvirus 8 genes on NF-kappaB activation. J Virol 83:2563–2574

11. Webb BL, Diaz B, Martin GS, Lai F (2003) A reporter system for reverse transfection cell arrays. J Biomol Screen 8:620–623

12. Tian L, Wang P, Guo J, Wang X, Deng W, Zhang C, Fu D, Gao X, Shi T, Ma D (2007) Screening for novel human genes associated with CRE pathway activation with cell microarray. Genomics 90:28–34

13. Silva JM, Mizuno H, Brady A, Lucito R, Hannon GJ (2004) RNA interference microarrays: high-throughput loss-of-function genetics in mammalian cells. Proc Natl Acad Sci U S A 101:6548–6552

14. Bailey SN, Ali SM, Carpenter AE, Higgins CO, Sabatini DM (2006) Microarrays of lentiviruses for gene function screens in immortalized and primary cells. Nat Methods 3:117–122

15. Neumann B, Held M, Liebel U, Erfle H, Rogers P, Pepperkok R, Ellenberg J (2006) High-throughput RNAi screening by time-lapse imaging of live human cells. Nat Methods 3:385–390

16. Neumann B, Walter T, Heriche JK, Bulkescher J, Erfle H, Conrad C, Rogers P, Poser I, Held M, Liebel U, Cetin C, Sieckmann F, Pau G, Kabbe R, Wunsche A, Satagopam V, Schmitz MH, Chapuis C, Gerlich DW, Schneider R, Eils R, Huber W, Peters JM, Hyman AA, Durbin R, Pepperkok R, Ellenberg J (2010) Phenotypic profiling of the human genome by time-lapse microscopy reveals cell division genes. Nature 464:721–727

17. Kato K, Umezawa K, Miyake M, Miyake J, Nagamune T (2004) Transfection microarray of nonadherent cells on an oleyl poly(ethylene glycol) ether-modified glass slide. Biotechniques 37:444–448, 450, 452.

18. Ji Q, Yamazaki T, Hanagata N, Lee MV, Hill JP, Ariga K (2012) Silica-based gene reverse transfection: an upright nanosheet network for promoted DNA delivery to cells. Chem Commun (Camb) 48:8496–8498

19. Wood KC, Konieczkowski DJ, Johannessen CM, Boehm JS, Tamayo P, Botvinnik OB, Mesirov JP, Hahn WC, Root DE, Garraway LA, Sabatini DM (2012) MicroSCALE screening reveals genetic modifiers of therapeutic response in melanoma. Sci Signal 5:rs4

20. Genovesio A, Giardini MA, Kwon YJ, de Macedo Dossin F, Choi SY, Kim NY, Kim HC, Jung SY, Schenkman S, Almeida IC, Emans N, Freitas-Junior LH (2011) Visual genome-wide RNAi screening to identify human host factors required for Trypanosoma cruzi infection. PLoS One 6:e19733

21. Genovesio A, Kwon YJ, Windisch MP, Kim NY, Choi SY, Kim HC, Jung S, Mammano F, Perrin V, Boese AS, Casartelli N, Schwartz O, Nehrbass U, Emans N (2011) Automated genome-wide visual profiling of cellular proteins involved in HIV infection. J Biomol Screen 16:945–958

22. Simpson JC, Cetin C, Erfle H, Joggerst B, Liebel U, Ellenberg J, Pepperkok R (2007) An RNAi screening platform to identify secretion machinery in mammalian cells. J Biotechnol 129:352–365

23. Simpson JC, Joggerst B, Laketa V, Verissimo F, Cetin C, Erfle H, Bexiga MG, Singan VR, Heriche JK, Neumann B, Mateos A, Blake J, Bechtel S, Benes V, Wiemann S, Ellenberg J, Pepperkok R (2012) Genome-wide RNAi screening identifies human proteins with a regulatory function in the early secretory pathway. Nat Cell Biol 14:764–774

24. Wheeler DB, Bailey SN, Guertin DA, Carpenter AE, Higgins CO, Sabatini DM (2004) RNAi living-cell microarrays for loss-of-function screens in Drosophila melanogaster cells. Nat Methods 1:127–132

25. Lindquist RA, Ottina KA, Wheeler DB, Hsu PP, Thoreen CC, Guertin DA, Ali SM, Sengupta S, Shaul YD, Lamprecht MR, Madden KL, Papallo AR, Jones TR, Sabatini DM, Carpenter AE (2011) Genome-scale RNAi on living-cell microarrays identifies novel regulators of Drosophila melanogaster TORC1-S6K pathway signaling. Genome Res 21:433–446

26. Rajan S, Djambazian H, Dang HC, Sladek R, Hudson TJ (2011) The living microarray: a high-throughput platform for measuring transcription dynamics in single cells. BMC Genomics 12:115

27. Rantala JK, Makela R, Aaltola AR, Laasola P, Mpindi JP, Nees M, Saviranta P, Kallioniemi O (2011) A cell spot microarray method for production of high density siRNA transfection microarrays. BMC Genomics 12:162

28. Fengler S, Bastiaens PI, Grecco HE, Roda-Navarro P (2012) Optimizing cell arrays for accurate functional genomics. BMC Res Notes 5:358

29. Fredlund E, Staaf J, Rantala JK, Kallioniemi O, Borg A, Ringner M (2012) The gene expression landscape of breast cancer is shaped by tumor protein p53 status and epithelial-mesenchymal transition. Breast Cancer Res 14:R113

30. Cepeda D, Ng HF, Sharifi HR, Mahmoudi S, Cerrato VS, Fredlund E, Magnusson K, Nilsson H, Malyukova A, Rantala J, Klevebring D, Vinals F, Bhaskaran N, Zakaria SM, Rahmanto AS, Grotegut S, Nielsen ML, Szigyarto CA, Sun D, Lerner M, Navani S, Widschwendter M, Uhlen M, Jirstrom K, Ponten F, Wohlschlegel J, Grander D, Spruck C, Larsson LG, Sangfelt O (2013) CDK-mediated activation of the SCF(FBXO) (28) ubiquitin ligase promotes MYC-driven transcription and tumourigenesis and predicts poor survival in breast cancer. EMBO Mol Med 5:1067–1086

31. Cekaite L, Rantala JK, Bruun J, Guriby M, Agesen TH, Danielsen SA, Lind GE, Nesbakken A, Kallioniemi O, Lothe RA, Skotheim RI (2012) MiR-9, -31, and −182 deregulation promote proliferation and tumor cell survival in colon cancer. Neoplasia 14:868–879

32. Rantala JK, Pouwels J, Pellinen T, Veltel S, Laasola P, Mattila E, Potter CS, Duffy T, Sundberg JP, Kallioniemi O, Askari JA, Humphries MJ, Parsons M, Salmi M, Ivaska J (2011) SHARPIN is an endogenous inhibitor of beta1-integrin activation. Nat Cell Biol 13:1315–1324

33. Kuhn E, Naschberger E, Konrad A, Croner RS, Britzen-Laurent N, Jochmann R, Münstedt H, Stürzl M (2012) A novel chip-based parallel transfection assay to evaluate paracrine cell interactions. Lab Chip 12:1363–1372

Chapter 2
Review of siRNA/shRNA Applications in Cell-Based Microarrays

Frank Weise

Abstract This chapter reviews the development and the application of siRNA cell-based microarrays. Starting from reverse transfection as a common ground, siRNA cell-based microarrays quickly evolved into assays of very different formats, and now range from arrays containing chemically synthesized siRNA to viral arrays encoding short hairpin RNA (shRNA), which gives rise to siRNA upon cell transduction. They encompass arrays printed onto microscopic slides as well as arrays in 384-well plates, with microwell chips as a speciality format in between. In order to reflect this variety of technical and biological solutions, representatives of each format are discussed, highlighting the particular features and the steps taken towards optimization of the siRNA/shRNA transfer efficiency. The chapter concludes with a brief discussion of the general limitations of this technology.

Abbreviations

APS	Aminopropylsilane
CDK2	Cyclin-dependent kinase 2
cDNA	Complementary DNA
dsDNA	Double-stranded DNA
ECM	Extracellular matrix
eGFP	Enhanced green fluorescent protein
GAPDH	Glyceraldehyde-3-phosphate dehydrogenase
GFP	Green fluorescent protein
hCAR	Human coxsackievirus and adenovirus receptor

F. Weise (✉)
The Natural and Medical Sciences Institute (NMI), University of Tübingen,
Reutlingen, Germany
e-mail: Weise@nmi.de

E. Palmer, *Cell-Based Microarrays: Review of Applications,*
Developments and Technological Advances, SpringerBriefs in Cell Biology,
DOI 10.1007/978-1-4939-0594-2_2, © Ella Palmer 2014

hMSCs	Human mesenchymal stem cells
HUVEC	Human umbilical vein endothelial cells
INCENP	Inner centromere protein
MAP2	Microtubule-associated protein 2
mCAT-1	Murine cationic amino acid receptor 1
MOI	Multiplicity of infection
MoMuLV	Moloney's murine leukemia virus
MSCV	Murine stem cell virus
mTOR	Mammalian target of Rapamycin
MVP	Major vault protein
nts	Nucleotides
PKR	Protein kinase R
PLK1	Polo-like kinase 1
PLL	Poly-L-lysine
RFP	Red fluorescent protein
RNAi	RNA interference
shRNA	Short hairpin RNA
siRNA	Small interfering RNA
TUNEL	Terminal deoxynucleotidyl transferase dUTP nick end labelling
VSV-G	Vesicular stomatitis virus glycoprotein

2.1 History

Microarrays are generally defined as two-dimensional arrays on a solid substrate ("chip") that allow the parallel analysis of a large number of biological samples using high-throughput screening methods. For most applications, the substrate is transparent and therefore suited for microscopy. A search in the PubMed database reveals more than 55,000 publications dealing with microarrays. Although initially developed in the early 1980s for the characterization of antibodies [1], microarrays first became widely popular in the 1990s with the development of DNA microarrays to probe gene expression.

Cell-based microarrays were first conceived for the so-called reverse transfection of adherent cells using cDNAs: In their ground-breaking Nature publication more than a decade ago, Junaid Ziauddin and David M. Sabatini printed the cDNA encoding a green fluorescent protein (GFP) in an aqueous gelatin solution onto a glass slide prior to providing a lipid transfection reagent, and then seeded HEK293 cells, forming a confluent monolayer [2]. The cells that happened to grow on the cDNA spots took up and expressed the cDNA, resulting in a microarray of spatially confined spots of fluorescent cells in an otherwise non-fluorescent cell layer. The diameter of the spots was about 120–150 μm, encompassing approximately 30–80 cells. In this context, the term "reverse transfection" refers to the order of the procedural steps, inasmuch as—contrary to the classical approach—first the DNA becomes immobilized before the cells attach to the surface. Reverse transfection is also often referred to as solid-phase transfection, from which the "classical" transfection is distinguished as liquid-phase transfection.

This new method was quickly picked up in the research community and after the work of Tom Tuschl paved the way for the application of small interfering RNAs (siRNAs) to mammalian cells in the same month [3], it was transferred to the field of RNA interference (RNAi): Two years later, in October 2003, the group of Olli Kallioniemi demonstrated the possibility to achieve RNAi-mediated knock-down in a microarray format by targeting enhanced green fluorescent protein (eGFP) expression in HeLa cells [4]. To this end, they printed siRNAs together with the transfection agent Lipofectin (Life Technologies) and the matrix-forming Matrigel (BD Biosciences) onto a glass slide. The spot size they employed ranged from 100 to 700 μm. They established a kinetic of the RNAi, observing that the knock-down efficiency increased over a time window from 24 to 72 h. Within the range of siRNA doses tested (from 0.25 to 2.5 ng per spot), only a moderate increase in the efficiency was observed.

In the same month, Vivek Mittal's group at Cold Spring Harbor Laboratory published a fluorescence-based reporter system suitable to effectively identify potent RNAi probes in the context of a cellular microarray [5]: By expressing a chimeric construct consisting of the cDNA of eGFP fused to the cDNA of a target gene, they generated a reporter transcript that allowed for the validation of the efficiency of siRNAs against that target gene using eGFP fluorescence as a read-out. The charm of this approach lies in the fact that a robust and simple optical read-out could be applied to any target, thus avoiding the need of target-specific detection methods such as immunofluorescence microscopy. In order to more easily identify the spots of transfected cells, a marker cDNA encoding a red fluorescent protein (RFP) was printed together with the siRNAs and the reporter plasmid. The employment of the RFP-encoding cDNA served a second purpose, namely the establishment of an internal standard: This allowed for the correction of differences in transfection efficiencies across the microarray by normalizing the eGFP fluorescence to the RFP fluorescence. The spot size in this publication was slightly larger (approximately 500 μm in diameter), encompassing 300–500 cells. Consequently, higher amounts of DNA—up to 300 ng—were employed. With this spot size and a pitch of 750 μm, approximately 2,500 spots can be accommodated on an area of 2 cm × 5 cm on a glass slide.

2.2 Optimization and Protocols

A protocol for the generation of a 384-spot array using siRNA in a gelatin matrix is given by [6]. The spot diameter achieved is 400 μm, sufficient to accommodate 100–200 HeLa cells. The pitch (spot-to-spot distance) is 1,125 μm, rendering this approach suitable for glass bottom cell culture dishes such as provided by Lab-Tek (Nalge Nunc International) or by MatTek Corp.

Sucrose and fibronectin are recommended as additives: The former facilitates the actual printing process by minimizing evaporation of the stocks, which is important given that the printing process can last several hours, depending on the robotics employed. Furthermore, it increases the shelf life of stored arrays, presumably by preventing complete desiccation of the samples. Printed mixtures of nucleic acids in a gelatin matrix, together with the transfection reagent, can be stored for prolonged periods of time: Erfle et al. [7] report a storage time of up to 15 months.

The second additive, fibronectin, increases the adherence of cells to the spots, reducing their propensity to migrate away once they have been transfected. A further potential function of fibronectin was discussed by [8]: By comparing the cell morphology of human mesenchymal stem cells (hMSCs) in the absence and the presence of fibronectin, they observed that both the cells and the nuclei spread much earlier and to a larger extent when fibronectin was present, in conjunction with a rapid relocation of actin filaments. Most importantly, fibronectin did not interfere with the differentiation potential of the hMSCs.

In their study, Yoshikawa et al. [8] furthermore undertook a thorough evaluation of six different transfection reagents regarding their suitability for reverse transfection in a microarray format, namely Lipofectamine 2000 (Invitrogen), JetPEI (q-Bio Gene), TransFast (Promega), Effectene, PolyFect, and SuperFect (all three from Qiagen), albeit restricted to plasmid-based transfection. Using hMSCs as well as HEK293, HeLa, NIH-3T3, and HepG2 cells, they established optimized transfection protocols. Finally, they addressed the question of glass coatings, comparing uncoated surfaces to surfaces coated with poly-L-lysine (PLL) or surfaces coated with aminopropylsilane (APS): On PLL-coated slides, fibronectin reached a saturation in its capability to improve transfection efficiency at approximately 0.5 μg/mL, whereas on APS-coated slides, it steadily increased transfection efficiency up to a concentration of 1.33 μg/mL, the highest concentration tested. At their respective optimal fibronectin concentrations, the transfection efficiencies achieved using PLL-coated slides and APS-coated slides were comparable.

In a continuation of this study by the same group, the protocols for reverse transfection were extended to PC-12 cells, a rat pheochromocytoma cell line commonly used as a surrogate for neuronal cells [9]: Collagen IV and to a lesser extent collagen I or laminin were found to significantly increase the transfection efficiency, whereas fibronectin or PLL did not. Interestingly, the order of potency of the three proteins in increasing transfection efficiency remained the same for six different transfection reagents tested: the highest efficiencies (>60 %) were achieved combining collagen IV with either TransFectin (Bio-Rad) or Lipofectamine 2000 (Life Technologies). Thus, this work underscores the necessity to finely adjust the transfection protocol to the cell type under investigation in order to achieve an optimal experimental outcome.

A similar optimization strategy was later on proposed by the same group: Instead of relying on various extracellular matrix (ECM) components to serve as additives supporting transfection efficiency, Uchimura et al. [10] suggested to employ antibodies directed against the receptors for ECM components in their place. Of these, a monoclonal antibody directed against CD29 (integrin subunit beta 1) proved to be the most efficient within the scope of four human cell lines (HeLa, HepG2, NT2, and HCC1428) and two rat cell types (PC-12 and rat fibroblasts) tested: The transfection efficiency, as determined by eGFP fluorescence, on slides coated with anti-CD29 antibody was either similar to or higher than the efficiency observed using fibronectin of murine collagen IV. These results imply the possibility to take a shortcut in the optimization procedure at least for such cells that are known to express CD29.

In order to improve the density of the spots that can be arrayed, Rantala et al. [11] exchanged glass slides for untreated polystyrene microplates (Nalge Nunc International), which are subdivided into four rectangular wells. With a spot diameter of 200 μm and a pitch of 500 μm, in an area of 18×54 mm, 3,888 spots could be arrayed. The printing of the siRNA array was similar to other protocols—using Matrigel as a matrix and sucrose as an additive, but the preparation and seeding of the cells was optimized to prevent adherence outside of the spots: For seeding, cells were detached using a nonenzymatic reagent (HyClone™ HyQTase, Thermo Scientific), and 2.5×10^6 cells in a volume of 4.5 mL of medium were added to a corner of a rectangular well, whilst avoiding direct application of the cells to the printed array. After an adherence time of 15 min, cells that had not adhered were washed off. This approach resulted in a lower number of cells per spot than usual: For instance, after 48 h of cultivation, 51 ± 3 PC-3 cells were found on a spot with a diameter of 200 μm, and 151 ± 3 cells on a 400 μm spot. Using primary human prostate stromal cells, the counts were even lower, as 21 ± 6 cells were found on a 200 μm spot. As a trade-off, the area outside of the spots was clear of cells, facilitating automated imaging and analysis. For 85 out of 92 cell types tested, this approach of spatially restricted adhesion was successful, indicating its suitability for most applications.

It is worth noting that in recent years the reverse transfection technology has spread from microarrays to multiwell plates. Note that in the multiwell format, the spot diameter is identical with the well diameter. Furthermore, the siRNA/gelatin matrix is not physically printed onto the surface, but is formed upon desiccation under vacuum from a larger volume. The suitability of this approach was demonstrated by knocking down the expression of cyclin-dependent kinase 2 (CDK2) in 15 human cell types, ranging from HEK293 and HeLa to mesenchymal stem cells [12]: In all but two lines (HCC1428 and AU565), a knock-down efficiency >80 % could be achieved using Lipofectamine200 as a transfection reagent and dextran and polyvinyl alcohol as additives.

Protocols for reverse transfection of siRNA, printed either in a microarray format or in a multiwell format, are abounding: For printing in a microarray format, the reader might consider [13] or [7], generating 450-μm spots. For reverse transfection using 96-well plates or 384-well plates consider [14] and [15]. Both the microarray and the multiwell format are considered in [16].

2.3 Microwell Arrays

In contrast to methodologies that generate an array of siRNAs on the plane surface of a chip, three groups have followed the approach to generate arrays of microwells on a chip. Their technologies constitute a hitherto missing link between the "classical" microarray of spots and the larger arrays of the 96/384/1536 microwell plate formats, combining the advantages of the former—high density and long-term storage, with the physical confinement of the latter.

In 2009, Reyman et al. demonstrated the feasibility of producing a 9,216 microwell array by photolithography, spreading a photoresist (AZ-Lack, Allresist, Germany) on a glass substrate and subsequently exposing it to UV light through an array-patterned mask. The unexposed photoresist was removed and replaced by titanium in an evaporation process under vacuum, after which the exposed photoresist was removed to reveal cavities with a diameter of 400 μm. By addressing two endogenously expressed genes, inner centromere protein (INCENP) and polo-like kinase-1 (PLK1) in HeLa cells, the group was able to reproduce the previously described phenotypes, namely multilobing of the cell nuclei in the case of INCENP knock-down and cell cycle arrest in the case of PLK1 knock-down.

Likewise by using photoresist film etching, Chen et al. [17] set up arrays from 2,666 wells to 65,800 wells. Whereas the former still provided room for 350–500 cells per cavity, in the latter the number of cells was about 30–50, clearly approaching the upper limit of wells desirable in such a format without compromising the statistical evaluation of the cellular response. For the 2,666 well format, 5–50 ng of siRNA in 3 μL of RNAiMAX (Invitrogen) were mixed with 11 μL of OptiMEM (Invitrogen), supplied with 0.2 M sucrose for simplifying printing, resulting in a volume of approximately 5 nL/spot. To these, approximately 4.6 μL of a HeLa cell suspension at 8.3×10^4 cells/ml were added. The diameter of the cavities was 500 μm, with a pitch of 100 μm. Using a cell line stably expressing an eGFP-DsRed fusion protein, the group showed a reduction in DsRed fluorescence after 72 h showing a dependence on the final concentration of the siRNA employed in the microcavities, ranging from 0.25 to 2.5 ng/μL.

A group at the Berlin Max Planck Institute developed a technology that allowed them to generate a reusable "carpet" with an array of wells on a chip [18]: To a standard glass slide, they applied a Teflon® master containing an array of 5×5 or 9×9 protuberances, respectively, and added a mixture of a silicone elastomer and its corresponding curing agent (SYLGARD® 184 silicone elastomer kit, Dow Corning). Upon hardening of the elastomer mixture and removal of the master, an array of wells was obtained sufficient to accommodate 500–1,000 cells in the 9×9 format (well surface area 1.23 mm^2), or 2,000–3,000 cells in the 5×5 format (well surface area 3.14 mm^2).

For the actual siRNA transfection, siRNAs at a concentration of 200 nM were pipetted into the cavities, followed by 3 μL of the pre-incubated and diluted transfection reagent (Dharmafect 1, diluted 1:100 in RPMI cell culture medium and incubated for 10 min; Thermo Scientific). After another 20 min of incubation, the target cells were seeded into the cavities. Targeting Aurora A in U-2 OS cells by siRNA, a drop in the number of mitotic cells by a factor of six (from 1.2 % down to 0.2 %) was observed 72 h after transfection, in line with the growth inhibition and apoptosis known to be the consequence of Aurora A knock-down.

2.4 shRNA-Mediated Knock-Down

The application of siRNAs as immediate effectors of knock-down—on a microarray or otherwise—is prone to dilution effects in proliferating cells, as the effective siRNA concentration per cell becomes reduced upon mitosis. A work-around for this problem lies in the constitutive replenishment of siRNAs by the cell, which is accomplished by the delivery of DNA coding for short hairpin RNAs (shRNAs). This approach was independently developed by Greg Hannon's group [19] and Reuven Agami's group [20], enabling the establishment of stable knock-down in mammalian cells: Both groups made use of RNA polymerase III promoters, the transcripts of which have defined start and end points. The transcripts described in these works contained a 26–29 nucleotides (nt) or a 19 nt long sequence, respectively, each specific to a target gene, followed by a 8–9 nt long spacer and the reverse complement of the specific sequence, which allows the RNA strand to fold back forming a hairpin structure, which is the reason why this structure was termed short hairpin RNA.

In 2004, Greg Hannon's group undertook the first steps towards the employment of shRNA-based microarrays in screening [21]: drawing upon 30 putative hits obtained from an earlier screen performed in the 96-well plate format, they employed a chimeric reporter construct containing a GFP coupled to a signal sequence mediating rapid proteasomal degradation (ZsProSensor-1, Clontech) to identify knock-down constructs that target proteasomal proteins: Cells on spots containing such shRNAs showed a gradual increase in green fluorescence intensity due to accumulation of the chimeric protein, reaching a maximum after 60–72 h.

Whereas in these initial works the application of shRNAs was restricted to plasmids, it became soon adapted to viral vectors: In December 2002, Patrick Aebischer's group reported a lentiviral vector expressing an shRNA directed against eGFP under the control of the human H1 promoter [22]. In their work, they showed that lentivirus-mediated shRNA-driven knock-down resulted in an approximately 30-fold reduction of the fluorescence signal in stably eGFP-expressing recombinant HEK293 cells, which persisted as long as 25 days.

In the following sections, we shall take a more detailed look at the application of retroviruses and lentiviruses as well as adenoviruses as means of delivery of shRNA into human cells.

2.4.1 Retro- and Lentivirus

The most commonly used retroviral systems are either based on Moloney's murine leukemia virus (MoMuLV), such as Clontech's RevTet™ system, or on murine stem cell virus (MSCV), such as Open Biosystem's Expression Arrest™ system. The entry of the MoMuLV virus into a target cell is mediated by the binding of its envelope glycoprotein gp70 to its cognate receptor, the murine cationic amino acid receptor 1 (mCAT-1). These viruses are ecotropic, i.e., their tropism is restricted to murine cells. For the infection of human cells, amphotropic variants are used, which bind to sodium-dependent phosphate symporters (e.g., PiT2; for details, see [23]). Such viruses have to be handled under biosafety level 2 laboratory conditions.

Using a spot-size of 500–800 μm in diameter, Carbone et al. [24] printed approximately 3,000 amphotropic biotinylated retroviral particles per spot on a nanostructured titania surface deposited using supersonic beam deposition of TiO_x, which was functionalized with streptavidin. Thus, the interaction between streptavidin and biotin was used to anchor the retrovirus to the spot. Depending on the cell type used, 200–400 cells were seeded on each spot. Approximately 800 spots could be accommodated on a standard microscopic slide when using a pitch of 1.5 mm. Addressing the expression of p53 in U-2 OS cells, a knock-down efficiency of >80 % was observed in more than a third of the transduced cells, and a knock-down efficiency at least 50 % was found in 75 % of the transduced cells, as was judged by immunofluorescence microscopy.

In the last years, lentiviruses became more popular than retroviruses as conveyors of genetic information. One advantage lentiviruses have over retroviruses is that they can integrate into the genome of nondividing cells, whereas retrovirus can only do so when the nuclear envelope is broken down during mitosis. Most commercially available lentivirus systems are based on HIV-1, but include the open reading frame of the vesicular stomatitis virus glycoprotein (VSV-G) in order to pseudotype the resulting recombinant viruses, thereby broadening their tropism (for a review, see [25]. Additionally, the VSV-G increases the stability of the lentivirus, rendering it more user-friendly. Like amphotropic retroviruses, lentiviruses have to be handled under biosafety level 2 laboratory conditions.

In February 2006, David M. Sabatini's group demonstrated the feasibility of lentivirus-infected cell microarrays [26]. Using glass slides coated with γ-amino propyl silane (UltraGAPS, Corning), they printed 3.9 nL of viral suspensions with a titre of 1×10^9 infectious units/ml (i.e., 3,900 infectious units/spot), giving rise to spots of 200–300 μm in diameter. When these slides were seeded with HeLa cells, approximately 150 cells came to lie over any viral spot area. Although the entire slide was covered with cells, the infection was locally confined to the cells located on the viral spots. Moreover, by alternatingly printing lentivirus-encoding GFP and lentivirus-encoding shRNA directed against human lamin A/C, and measuring GFP fluorescence as well as detecting lamin A/C by immunofluorescence microscopy, no significant cross-contamination was observed: GFP fluorescence was confined to cells located on spots of GFP-encoding lentiviruses, and knock-down of lamin A/C, which reduced the expression to approximately 20 % and was restricted to cells located on spots of shRNA-encoding lentiviruses.

Using GFP-encoding lentiviruses, Bailey et al. extended their work to both dividing and post-mitotic primary cells: Primary human BJ fibroblasts served as a representative of primary dividing cells, whereas murine bone marrow-derived dendritic cells served as a representative of nondividing cells. A higher degree of complexity was achieved in an experiment conducted on HeLa cells, which were seeded on arrays printed with lentivirus-encoding shRNA directed against lamin A/C (as a negative control) or against mammalian target of rapamycin (mTOR), a kinase involved in signal transduction cascades regulating cell size and proliferation. Cells located on spots of viruses encoding the shRNA against mTOR showed not only a reduction in the amount of phosphorylated protein S6, a downstream target in the mTOR-signaling cascade, as was determined by immunofluorescence, but also were smaller in size and fewer in number than were cells located on control spots.

Several research institutions joined their efforts to provide the scientific community with validated shRNA sequences in lentiviral vectors, forming a public–private consortium (The RNAi Consortium; http://www.broadinstitute.org/rnai/trc). The suitability of the initial portion of the shRNA library thus generated was confirmed in an initial screen performed in a 384-well plate format, comprising 4,903 lentiviruses [27]. This library is commercially available through the private consortium partners (Sigma Aldrich; Thermo Fisher).

2.4.2 Adenovirus

It should be noted that most of the adenovirus work has been performed using the serotype 5 (Ad5), which is also used in most commercially available expression systems. In contrast to retro-/lentiviruses, adenoviruses do not integrate into their host's genome. Furthermore, it is to be considered that for an efficient infection, adenovirus require the presence of the human coxsackievirus and adenovirus receptor (hCAR) on the cell surface, excluding, among others, skeletal muscle cells, endothelial cells, and hematopoietic cells from adenovirus-mediated transfer (reviewed in [28]). However, the adenoviral transfection efficiency can be improved by various compounds, of which the non-liposomal transfection reagent FuGENE6™ (Roche) is the least toxic [29]. As the retro- and lentiviruses discussed above, adenoviruses have to be handled under biosafety level 2 laboratory conditions.

In 2003, Zhao et al. [30] established adenovirus-mediated knock-down using an shRNA transcript driven by the H1 promoter in HeLa cells. In their work on murine jejunal *Tunica muscularis* cells, Ro et al. [29] performed a comparative study of three RNA polymerase III promoters in the same adenoviral background: Using Stratagene's AdEasy™ Adenoviral Vector System, they introduced shRNA sequences directed against glyceraldehyde-3-phosphate dehydrogenase (GAPDH) under control of either the murine or the human U6 small nuclear RNA promoter, or the human H1 promoter, into the target cells. The knock-down efficiencies observed using Western blot analysis were highly similar, namely reductions by 80 ± 6 % for the murine U6 promoter, 75 ± 8 % for the human U6 promoter, and 88 ± 4 % for the human H1 promoter, indicating that, at least in that particular cell system, the choice of the promoter does not exert a strong effect on the knock-down efficiencies achievable.

The first functional screening using an adenoviral array was described by Michiels et al. [31]: Using a 96-well plate format, they generated adenovirus suspensions with a titre of 2.5×10^9 viral particles per ml, harboring a cDNA library. These were transferred to another 96-well plate into which the target cells, hMSCs, had been seeded previously, which became transfected at an average multiplicity of infection (MOI) of 2,500.

In 2008, Oehmig et al. [32] applied the principle of virus-mediated transfer as described by Bailey et al. [26] to an array of shRNA-encoding adenoviruses: In contrast to the earlier work of David M. Sabatini's group, they printed the adenovirus on a glass slide but then blocked the surface outside the spots, so that cells could become adherent only on the adenoviral spots, which greatly simplified the identification of viral spots without the need of additional fluorescent markers. To this end, the glass slides were coated with nitrocellulose prior to printing the viruses, and blocked by the application of StabilGuard (SG01, SurModics). With a diameter of 80 μm, the spots were a smaller size than commonly printed. Using a combination of Invitrogen's U6-promoter-driven shRNA expression system in the context of a site-specific recombination system (BLOCK-iT™ U6 RNAi Entry Vector Kit) for easier shuttling and Stratagene's AdEasy™ Adenoviral Vector, Oehmig et al. functionally analyzed the knock-down of a token panel of seven human protein kinases on the human osteosarcoma cell line U-2 OS. The panel had been selected to include known regulators of apoptosis, and the shRNA sequences had been pre-validated to achieve a knock-down by at least 60 %. In the adenoviral microarray, four out of the seven knock-down sequences exerted a phenotypic effect, namely the induction of apoptosis, as determined independently by a fluorescence-based TUNEL assay and by immunofluorescence microscopy detecting cleaved caspase 3, a hallmark of apoptosis. In a last step, the adenoviral shRNA array was applied to human umbilical vein endothelial cells (HUVEC), thereby extending the range from transformed cell lines to primary cells: Again, four out of the seven knock-down sequences (although not the same four, reflecting differences in the cell biology of the target cells) proved to be effective in eliciting apoptosis.

A protocol for the generation and application microarrays of shRNA-encoding adenovirus is given in [33].

2.5   shRNA Sequence

For the selection of siRNA sequences, a couple of empirical rules have been established, which resulted in the set up of several public-domain servers, such as Wadsworth's Sfold server (http://sfold.wadsworth.org/cgi-bin/index.pl; [34] or Whithead's siRNA selection program (http://sirna.wi.mit.edu; [35], the discussion of which are outside the scope of this review. Suffice to say that the algorithms employed for the prediction of siRNA sequences are not suited to predict the efficiency of shRNA sequences to the same degree, as a comparative study revealed [36].

Considering that microarrays lend themselves to the screening of rare or limited samples, we take a short look at the generation of shRNA libraries generated from cDNA, as these take a great step forward towards personalized libraries: In 2004, three publications appeared which focus on the enzymatic generation of shRNA libraries [37–39], followed by another work in 2005 [40], and two more in 2006 [41, 42]. All these approaches make use of the restriction enzyme MmeI, which cleaves DNA 20–21 bp away from its recognition site, thus giving rise to 21-nt-long stretches of DNA, which contain the target strand as well as its reverse complement, already in the length preferred for efficient siRNAs. The approaches taken in these six publications differ in the types of adapters that are ligated to the dsDNA, which at some stage contain a pre-formed hairpin structure and thus upon ligation render the construct reminiscent of the shRNA (though still being a DNA molecule), and they also differ in the polymerases employed in order to amplify the DNA hairpin for subsequent cloning into suitable vectors. Nonetheless, at the end, all approaches result in a stretch of DNA encoding the shRNA molecule.

It should be noted that generating an shRNA library from a cDNA as such would result in a biased representation of the genome, as some transcripts are considerably more abundant than others. Normalization techniques have been developed to overcome the unbalanced representation, favoring the representation of less abundant transcripts over the more abundant one. For an overview of such approaches, consider [43] and references therein.

2.6 Variation of siRNA Delivery

All the methods discussed so far rely upon protein-based matrices such as gelatin, Matrigel, fibronectin, or others to locally confine either the siRNA or the viruses on the spots.

An interesting variation of the theme was presented by the work of Zhao et al. [44], which employed an shRNA library synthesized in an attenuated *E. coli* strain and made use of integrin-receptor-based bacterial invasion of HeLa cells to deliver the shRNAs into the mammalian cells. At an MOI of 100, approximately 80 % of the HeLa cells were invaded, albeit at the price of some cell death. A reduction in the MOI to 20 was accompanied with a reduction of the gene transfer efficiency to 30 %. The shRNA delivered in this manner was efficient in knocking-down the expression of an endogenous gene, major vault protein (MVP), by approximately 50 % as judged by Western blot analysis. Remarkably, the bacterial invasion did not cause a detectable interferon response in the HeLa cells, as was established by the analysis of the levels of phosphorylated protein kinase R (PKR), which is induced by interferon and becomes activated upon binding to double-stranded RNA. This approach was successfully employed in a wider shRNA screen, in which an enzymatically generated shRNA library against MVP was delivered in a 96-well format to HeLa cells, resulting in the identification of five potent shRNA sequences. Although not yet in a microarray format—the read out was based on the luciferase activity of a luciferase-MVP chimeric transcript—this method bears the potential for the application in a microarray format at least in those cases in which no interference of the bacterial invasion with cellular processes is anticipated.

In two studies published in 2010 and 2012, Hongkun Park and collaborators developed a novel approach addressing cells typically hard to transfect, such as primary neurons and cells of the immune system [45, 46]: They generated an array of silicon nanowires (approximately 1–3 μm in length and 150 nm in diameter, with densities ranging from 0.15 to 1 per μm^2), which were subsequently functionalized using aminosilane to allow for the non-covalent binding of siRNA molecules. More than 95 % of the cells seeded on this spiked surface became penetrated by the nanowires within an hour. Temporarily, the cells developed irregular contours, but returned to normal morphology within a couple of hours. Notably, immune cells tested such as murine B cells, dendritic cells, and macrophages showed low basal expression levels of the inflammatory marker tumor necrosis factor α, while exhibiting normal inflammatory responses when stimulated. These findings were taken to indicate that the impalement caused by the nanowires did not elicit any significant immune response, which is presumably due to the fact that nanowire penetration directly delivers the cargo into the cytosol, thus bypassing the endosomal pathway and the immune checkpoints associated with it.

This approach lends itself to knock-down studies in a microarray format: In a feasibility study addressing vimentin expression in HeLa S3 cells, Shalek et al. [45] observed a significant reduction in vimentin expression as monitored by immunofluorescence microscopy.

2.7 Concluding Remarks

The chip-based microarrays discussed in this chapter mostly result in a monolayer of cells, of which only those cells that happen to be seeded on a spot take up the genetic information. Thus, the cells are free to leave "their" spot and move around later on. This lack of confinement might be problematic for some sorts of assays, and can be either overcome by blocking the area outside the spots to cell adherence, as shown by Oehmig et al. [32], by a combination of hydrophobic surfaces and shortened time for adherence [11], or by the generation of microwells [17, 18, 47], which puts up a physical barrier around the cells.

Presumably due to the adaptation of the commercially available robotics platforms to the 96/384/1536 format, many published high-throughput/high content library screens making use of cell-based reverse transfection arrays were performed in a 384-well plate format, examples of which are given in [48] and [49] for siRNA-based arrays, and in [27] for lentiviral arrays.

It should be noted that, for most cell systems, the application of the microarray technology is restricted to experiments in which a detectable effect is expected to occur within approximately 72 h, given that cells would become overgrown if incubated for longer periods of time. An exception to this limitation are post-mitotic cells such as neurons, for which Krichevsky and Kosik [50] demonstrated the applicability of siRNA-mediated RNAi: By using siRNAs directed against an endogenously expressed gene (microtubule-associated protein 2, MAP2), they achieved an approximately fourfold reduction in the protein level as judged by immunofluorescence microscopy performed 48–68 h after transfection, affecting approximately 70–80 % of the neurons.

Given the rather short experimental time window of the cell-based microarrays, the advantage of employing shRNAs over siRNAs, namely the generation of stable siRNA expression, does not appear to bear fruit. However, for any downstream characterization of hits picked during the screening, it pays off to have identified efficient shRNA sequences given that the efficiency of any given sequence when applied in an shRNA-based knock-down is different from the one achieved employing it in an siRNA-based approach [36].

References

1. Chang TW (1983) Binding of cells to matrixes of distinct antibodies coated on solid surface. J Immunol Methods 65(1–2):217–223
2. Ziauddin J, Sabatini DM (2001) Microarrays of cells expressing defined cDNAs. Nature 411(6833):107–110
3. Elbashir SM, Harborth J, Lendeckel W, Yalcin A, Weber K, Tuschl T (2001) Duplexes of 21-nucleotide RNAs mediate RNA interference in cultured mammalian cells. Nature 411(6836):494–498
4. Mousses S, Caplen NJ, Cornelison R, Weaver D, Basik M, Hautaniemi S, Elkahloun AG, Lotufo RA, Choudary A, Dougherty ER, Suh E, Kallioniemi O (2003) RNAi microarray analysis in cultured mammalian cells. Genome Res 13(10):2341–2347
5. Kumar R, Conklin DS, Mittal V (2003) High-throughput selection of effective RNAi probes for gene silencing. Genome Res 13(10):2333–2340
6. Erfle H, Pepperkok R (2005) Arrays of transfected mammalian cells for high content screening microscopy. Methods Enzymol 404:1–8
7. Erfle H, Neumann B, Liebel U, Rogers P, Held M, Walter T, Ellenberg J, Pepperkok R (2007) Reverse transfection on cell arrays for high content screening microscopy. Nat Protoc 2(2):392–399
8. Yoshikawa T, Uchimura E, Kishi M, Funeriu DP, Miyake M, Miyake J (2004) Transfection microarray of human mesenchymal stem cells and on-chip siRNA gene knockdown. J Control Release 96(2):227–232
9. Uchimura E, Yamada S, Uebersax L, Yoshikawa T, Matsumoto K, Kishi M, Funeriu DP, Miyake M, Miyake J (2005) On-chip transfection of PC12 cells based on the rational understanding of the role of ECM molecules: efficient, non-viral transfection of PC12 cells using collagen IV. Neurosci Lett 378(1):40–43
10. Uchimura E, Yamada S, Nomura T, Matsumoto K, Fujita S, Miyake M, Miyake J (2007) Reverse transfection using antibodies against a cell surface antigen in mammalian adherent cell lines. J Biosci Bioeng 104(2):152–155
11. Rantala JK, Mäkelä R, Aaltola AR, Laasola P, Mpindi JP, Nees M, Saviranta P, Kallioniemi O (2011) A cell spot microarray method for production of high density siRNA transfection microarrays. BMC Genomics 12:162. doi:10.1186/1471-2164-12-162
12. Fujita S, Ota E, Sasaki C, Takano K, Miyake M, Miyake J (2007) Highly efficient reverse transfection with siRNA in multiple wells of microtiter Plates. J Biosci Bioeng 104(4):329–333
13. Erfle H, Pepperkok R (2007) Production of siRNA- and cDNA-transfected cell arrays on non-coated chambered coverglass for high-content screening microscopy in living cells. Methods Mol Biol 360:155–161
14. Erfle H, Neumann B, Rogers P, Bulkescher J, Ellenberg J, Pepperkok R (2008) Work flow for multiplexing siRNA assays by solid-phase reverse transfection in multiwell Plates. J Biomol Screen 13(7):575–580. doi:10.1177/1087057108320133

15. Henderson MC, Azorsa DO (2013) High-throughput RNAi screening for the identification of novel targets. Methods Mol Biol 986:89–95. doi:10.1007/978-1-62703-311-4_6
16. Fujita S, Takano K, Ota E, Sano T, Yoshikawa T, Miyake M, Miyake J (2010) New methods for reverse transfection with siRNA from a solid surface. Methods Mol Biol 623:197–209. doi:10.1007/978-1-60761-588-0_13
17. Chen PC, Huang YY, Juang JL (2011) MEMS microwell and microcolumn arrays: novel methods for high-throughput cell-based assays. Lab Chip 11(21):3619–3625. doi:10.1039/c0lc00696c
18. Scholz AK, Klebl BM, Morkel M, Lehrach H, Dahl A, Lange BM (2010) A flexible multiwell format for immunofluorescence screening microscopy of small-molecule inhibitors. Assay Drug Dev Technol 8(5):571–580. doi:10.1089/adt.2009.0260
19. Paddison PJ, Caudy AA, Bernstein E, Hannon GJ, Conklin DS (2002) Short hairpin RNAs (shRNAs) induce sequence-specific silencing in mammalian cells. Genes Dev 16(8):948–958
20. Brummelkamp TR, Bernards R, Agami R (2002) A system for stable expression of short interfering RNAs in mammalian cells. Science 296(5567):550–553
21. Silva JM, Mizuno H, Brady A, Lucito R, Hannon GJ (2004) RNA interference microarrays: high-throughput loss-of-function genetics in mammalian cells. Proc Natl Acad Sci U S A 101(17):6548–6552
22. Abbas-Terki T, Blanco-Bose W, Déglon N, Pralong W, Aebischer P (2002) Lentiviral-mediated RNA interference. Hum Gene Ther 13(18):2197–2201
23. Beer C, Andersen DS, Rojek A, Pedersen L (2005) Caveola-dependent endocytic entry of amphotropic murine leukemia virus. J Virol 79(16):10776–10787
24. Carbone R, Giorgetti L, Zanardi A, Marangi I, Chierici E, Bongiorno G, Fiorentini F, Faretta M, Piseri P, Pelicci PG, Milani P (2007) Retroviral microarray-based platform on nanostructured TiO$_2$ for functional genomics and drug discovery. Biomaterials 28(13):2244–2253
25. Cronin J, Zhang XY, Reiser J (2005) Altering the tropism of lentiviral vectors through pseudo-typing. Curr Gene Ther 5(4):387–398
26. Bailey SN, Ali SM, Carpenter AE, Higgins CO, Sabatini DM (2006) Microarrays of lentiviruses for gene function screens in immortalized and primary cells. Nat Methods 3(2):117–122
27. Moffat J, Grueneberg DA, Yang X, Kim SY, Kloepfer AM, Hinkle G, Piqani B, Eisenhaure TM, Luo B, Grenier JK, Carpenter AE, Foo SY, Stewart SA, Stockwell BR, Hacohen N, Hahn WC, Lander ES, Sabatini DM, Root DE (2006) A lentiviral RNAi library for human and mouse genes applied to an arrayed viral high-content screen. Cell 124(6):1283–1298
28. Volpers C, Kochanek S (2004) Adenoviral vectors for gene transfer and therapy. J Gene Med 6(Suppl 1):S164–S171
29. Ro S, Hwang SJ, Ordög T, Sanders KM (2005) Adenovirus-based short hairpin RNA vectors containing an EGFP marker and mouse U6, human H1, or human U6 promoter. Biotechniques 38(4):625–627
30. Zhao LJ, Jian H, Zhu H (2003) Specific gene inhibition by adenovirus-mediated expression of small interfering RNA. Gene 316:137–141
31. Michiels F, van Es H, van Rompaey L, Merchiers P, Francken B, Pittois K, van der Schueren J, Brys R, Vandersmissen J, Beirinckx F, Herman S, Dokic K, Klaassen H, Narinx E, Hagers A, Laenen W, Piest I, Pavliska H, Rombout Y, Langemeijer E, Ma L, Schipper C, Raeymaeker MD, Schweicher S, Jans M, van Beeck K, Tsang IR, van de Stolpe O, Tomme P, Arts GJ, Donker J (2002) Arrayed adenoviral expression libraries for functional screening. Nat Biotechnol 20(11):1154–1157
32. Oehmig A, Klotzbücher A, Thomas M, Weise F, Hagner U, Brundiers R, Waldherr D, Lingnau A, Knappik A, Kubbutat MH, Joos TO, Volkmer H (2008) A novel reverse transduction adenoviral array for the functional analysis of shRNA libraries. BMC Genomics 9:441. doi:10.1186/1471-2164-9-441
33. Volkmer H, Weise F (2011) Cell-based microarrays of infectious adenovirus encoding short hairpin RNA (shRNA). Methods Mol Biol 706:97–106. doi:10.1007/978-1-61737-970-3_8
34. Ding Y, Chan CY, Lawrence CE (2004) Sfold web server for statistical folding and rational design of nucleic acids. Nucleic Acids Res 32(Web Server issue):W135–W141

35. Yuan B, Latek R, Hossbach M, Tuschl T, Lewitter F (2004) siRNA Selection Server: an automated siRNA oligonucleotide prediction server. Nucleic Acids Res 32(Web Server issue):W130–W134

36. Taxman DJ, Livingstone LR, Zhang J, Conti BJ, Iocca HA, Williams KL, Lich JD, Ting JP, Reed W (2006) Criteria for effective design, construction, and gene knockdown by shRNA vectors. BMC Biotechnol 6:7

37. Sen G, Wehrman TS, Myers JW, Blau HM (2004) Restriction enzyme-generated siRNA (REGS) vectors and libraries. Nat Genet 36(2):183–189

38. Luo B, Heard AD, Lodish HF (2004) Small interfering RNA production by enzymatic engineering of DNA (SPEED). Proc Natl Acad Sci U S A 101(15):5494–5499

39. Shirane D, Sugao K, Namiki S, Tanabe M, Iino M, Hirose K (2004) Enzymatic production of RNAi libraries from cDNAs. Nat Genet 36(2):190–196

40. Dinh A, Mo YY (2005) Alternative approach to generate shRNA from cDNA. Biotechniques 38(4):629–632

41. Du C, Ge B, Liu Z, Fu K, Chan WC, McKeithan TW (2006) PCR-based generation of shRNA libraries from cDNAs. BMC Biotechnol 6:28

42. Xu L, Li J, Liu L, Lu L, Gao J, Li X (2007) Construction of equalized short hairpin RNA library from human brain cDNA. J Biotechnol 128(3):477–485

43. Zhulidov PA, Bogdanova EA, Shcheglov AS, Vagner LL, Khaspekov GL, Kozhemyako VB, Matz MV, Meleshkevitch E, Moroz LL, Lukyanov SA, Shagin DA (2004) Simple cDNA normalization using kamchatka crab duplex-specific nuclease. Nucleic Acids Res 32(3):e37

44. Zhao HF, L'Abbé D, Jolicoeur N, Wu M, Li Z, Yu Z, Shen SH (2005) High-throughput screening of effective siRNAs from RNAi libraries delivered via bacterial invasion. Nat Methods 2(12):967–973

45. Shalek AK, Gaublomme JT, Wang L, Yosef N, Chevrier N, Andersen MS, Robinson JT, Pochet N, Neuberg D, Gertner RS, Amit I, Brown JR, Hacohen N, Regev A, Wu CJ, Park H (2012) Nanowire-mediated delivery enables functional interrogation of primary immune cells: application to the analysis of chronic lymphocytic leukemia. Nano Lett 12(12):6498–6504. doi:10.1021/nl3042917

46. Shalek AK, Robinson JT, Karp ES, Lee JS, Ahn DR, Yoon MH, Sutton A, Jorgolli M, Gertner RS, Gujral TS, MacBeath G, Yang EG, Park H (2010) Vertical silicon nanowires as a universal platform for delivering biomolecules into living cells. Proc Natl Acad Sci U S A 107(5):1870–1875. doi:10.1073/pnas.0909350107

47. Reymann J, Beil N, Beneke J, Kaletta PP, Burkert K, Erfle H (2009) Next-generation 9216-microwell cell arrays for high-content screening microscopy. Biotechniques 47(4): 877–878. doi:10.2144/000113251

48. Henderson MC, Gonzales IM, Arora S, Choudhary A, Trent JM, Von Hoff DD, Mousses S, Azorsa DO (2011) High-throughput RNAi screening identifies a role for TNK1 in growth and survival of pancreatic cancer cells. Mol Cancer Res 9(6):724–732. doi:10.1158/1541-7786. MCR-10-0436

49. Azorsa DO, Gonzales IM, Basu GD, Choudhary A, Arora S, Bisanz KM, Kiefer JA, Henderson MC, Trent JM, Von Hoff DD, Mousses S (2009) Synthetic lethal RNAi screening identifies sensitizing targets for gemcitabine therapy in pancreatic cancer. J Transl Med 7:43. doi:10.1186/1479-5876-7-43

50. Krichevsky AM, Kosik KS (2002) RNAi functions in cultured mammalian neurons. Proc Natl Acad Sci U S A 99(18):11926–11929

Chapter 3
Antibody Microarrays for Cell-Based Assays: The Use of Micro-Arrayed Antibodies for Exploring Cell Surface Diversity or Whole Cell Functionality

Yoann Roupioz

Abstract Since the early 1980s, microarrays have gained increasing interest due to their tremendous field of applications. A large repertoire of biomolecules has been micro-arrayed for the specific and parallelized detection of targets contained in a single biological sample. Most individual binding events target soluble compounds although pioneering microarray developments have been undertaken with whole cell binding on micro-arrayed glass slides. In this review, we wish to focus on micro-arrayed antibodies and their use for cell-based analysis. We will illustrate the wide range of applications that can be explored by site-specifically immobilizing eukaryotic or prokaryotic cells on a surface.

Abbreviations

AML	Acute myeloid leukemia
CCM	Culture-capture-measure
CD	Clusters of differentiation
CFU	Colony forming units
CTC	Circulating tumor cells
DNA	Deoxynucleic acid
ELISA	Enzyme-linked immunosorbent assay
FACS	Fluorescence-assisted cell sorting
GNP	Gold nanoparticles
HAART	Highly active antiretroviral therapy
HIV	Human immunodeficiency virus

Y. Roupioz, Ph.D. (✉)
SPrAM, UMR 5819 CEA/CNRS/UJF-Grenoble 1, Institute for Nanosciences and Cryogeny,
CEA-Grenoble, 38054 Grenoble, France
e-mail: yoann.roupioz@cea.fr

E. Palmer, *Cell-Based Microarrays: Review of Applications,*
Developments and Technological Advances, SpringerBriefs in Cell Biology,
DOI 10.1007/978-1-4939-0594-2_3, © Ella Palmer 2014

ICS   Intra-cellular cytokine staining
IFNγ   Interferon gamma
IL   Interleukin
IS   Immunological synapses
MHC   Major histocompatibility complex
mRNA   Messenger ribonucleic acid
rRNA   Ribosomal ribonucleic acid
LOD   Limit of detection
PBMC   Peripheral blood mononuclear cell
PCR   Polymerase chain reaction
PDMS   Polydimethylsiloxane
RLS   Resonant light scattering
SECM   Scanning electrochemical microscopy
SLE   Systemic lupus erythematosus
SNP   Single nucleotide polymorphisms
SPR   Surface plasmon resonance
STEC   Shiga toxin-producing *E. coli*
TCR   T cell receptor
TNFα   Tumor necrosis factor

3.1 Introduction

Micro- and nanotechnologies have been acquiring increasing interest for more than two decades. This general enthusiasm is motivated by several reasons; the most important is undoubtedly due to the scale of these new generations of devices. By decreasing the size of experimental setups, one can reasonably expect the decrease in volume of samples required to run an analysis on a micro-system. This feature might be critical in biological or biomedical applications as, in most of them, access to large sample volumes might be challenging. Another important advantage in using micro- and nano-systems in biological studies is the decrease in time usually required for running multistep processes. This aspect is illustrated by recent advances in Laboratories-On-A-Chip (Lab-On-Chip), where integrated functions allow, for instance, the efficient analysis and sorting (i.e., physical separation) of rare circulating tumor cells (CTCs) carried in the blood stream [1]. Last but not least, another major benefit in using miniaturized devices is the possibility to run parallelized assays on a single support. This multiplexed analysis might be carried out according to two aspects: one is based on a series of samples that can be handled and analyzed in parallel, thanks to dedicated microfluidic architectures [2], while the other is focused on one single sample interrogated at several points throughout a single analysis. A typical example of this latter point is illustrated by microarrays where series of probes (DNA [3], proteins [4], peptides [5], carbohydrates [6], etc.) are immobilized on predetermined locations of a surface. Upon incubation with a sample, individual responses of each micro-arrayed probe are simultaneously analyzed and compared to control responses, thus producing a significant amount of information through a single assay.

So far, the most abundant applications of microarrays are issued from DNA microarrays dedicated to the probing of transcriptomes, genomes, and searches for single nucleotide polymorphisms (SNP). Pioneer work on DNA microarrays was led at Stanford University by Pat Brown's group which first published the use of DNA microarrays for gene expression level assessments [7]. Although tremendous interest arose from DNA microarray-based technologies, these miniaturized systems functionalized by biomolecular probes are not the oldest format of parallelized arrays. Indeed, besides Fodor's group which is often cited as the first example of micro-arrayed peptides [8], Ekin's lab proposed in the late 1980s the simultaneous arraying of antibodies for immunoassay applications [9, 10]. This original work was based on a "multi-analyte microspot" assay where series of antibodies were simultaneously immobilized on a solid support and then incubated with biological samples. Using such formats, both noncompetitive and competitive assays were led by combining secondary radio-labeled antibodies and labeled antigens, respectively [9, 11]. Data produced by these protein microarrays gave both relative occupancy and quantitative data on surface immobilized antibodies. Since then, protein microarrays met an increasing enthusiasm, giving rise to two major classes of protein microarrays: protein function arrays made of immobilized libraries of proteins for assessment of their function; and protein detection arrays which consist of immobilizing proteins with defined binding properties used for probing targets contained in a sample [12]. Antibody microarrays definitely fall into this latter class and are systematically used for antigen binding, followed by detection.

Soluble antigen detection using antibody microarrays [13] or microarrays designed for cell lysate and tissue extracts [14] will not be described here as this chapter is focused on cell-based assays led on antibody microarrays. By carefully looking at the literature, this latter category appears as the pioneer demonstration of the microarray concept ever described and illustrated in Chang's work [15], published in 1983, describing immobilization of anti-Lyt 2.1 and anti-Lyt 2.2 antibodies (1, 0.5, and 0.25 mm spots in diameter) on a solid support. Several hundreds of features (up to 400) were micro-arrayed on 1 cm^2 and the whole microarray was incubated with cell lines expressing either one or the other antigen on their cell membrane. After washing, AKR thymocytes expressing the Lyt 2.1 antigen and BALB/c mouse thymocytes expressing the Lyt 2.2 antigen specifically bound to the anti-Lyt 2.1 and anti-Lyt 2.2 spots, respectively. Individual cells bound to each feature were counted and the measured counts of cells were consistent with the spot area (more captured cells on larger spots), thus validating the surface antigen exploration using a solid support. This work appears to be the very first example of micro-arrayed biomolecules, and interestingly was dedicated to whole-cell analysis. More recently, several examples of cell microarrays have been published [16, 17], driven by an important demand for new devices specific to cell analysis [18]. Blood cells are of special interest due to the ease of collection by blood tests [19], although protein microarrays have also been described for phenotyping adhering cells (mesenchymal stem cells) [20] or for probing cell adhesion properties of hepatocytes [21]. Most blood cells show a weak ability for spontaneous binding onto a surface, which means that specific cell capture and analysis on a microarray implies specific cell surface interactions with functionalized spots. These cell surface recognition events are mostly ensured by recognition of clusters of differentiation (CDs) by specific antibodies as shown in the following examples. This strategy has been used for functionalizing patterns [22] or microstructures [23] with one type of antibody to specifically capture individual cells on defined locations. In both examples, the whole microarray is then turned into a matrix of identical cells enabling either easy readout for subsequent bioassays [22] or even sorting by dissociating pieces of the solid support bearing immobilized cells [23]. This chapter will focus on recent developments on protein microarrays grafted by a series of different antibodies.

3.2 Antibody Microarrays for Cell Typing

3.2.1 Phenotyping Blood Cells

CDs are surface motifs identified as specific to a cell type, sub-cell type, differentiation stage, or physiological process, etc. By immobilizing antibodies targeting a specific CD, one may expect to obtain some cell arrangement on two dimensions depending on their surface antigen. This strategy has been successfully used to distinguish undifferentiated from differentiated fetal neural cells [24]. In this work, authors showed that the number of differentiated neural cells was higher on anti-CD56 and anti-O4, lower on anti-CD57 and similar on anti-CD44, CD81, and anti-CD90 spots, by comparison to similar experiments done with undifferentiated cells. These data were confirmed by counter-staining of the captured cells by labeling of intracellular markers. Multiplexed information is then accessible by combining microarray binding position to cross-labeling specific to another antigen. This explains why this approach is sometimes referred as "solid-phase cytometry" [24].

From a quantitative point of view, absolute weighing of different cell populations is delicate as the cell capture is highly dependent on several factors such as the antibodies ability to bind to its antigen or the abundance of the antigen on the cell membrane. Nevertheless, semiquantitative data has been reported [24–26]. Some authors even showed how real-time monitoring of cell binding on an antibody microarray could yield information on diffusion coefficients for membrane antigens diffusing around the cells [26]. The measured values were consistent with diffusion coefficients measured by photo-bleaching experiments. Within one blood cell family, subsets of cells were also distinguished by micro-patterning anti-CD4 and anti-CD8 antibodies on a surface [25, 27]. Ratios of these T-lymphocyte cells were also confirmed by control flow cytometry experiments. The coupling of such microarrays with dedicated microfluidic systems showed remarkable results obtained by flowing fleshly drawn peripheral blood mononuclear cells (PBMCs) [27], which might be a useful tool for blood analysis in General Practitioner offices. Other screens were also done for exploring a larger repertoire of surface antigens. Ellmark et al. [28] arrayed

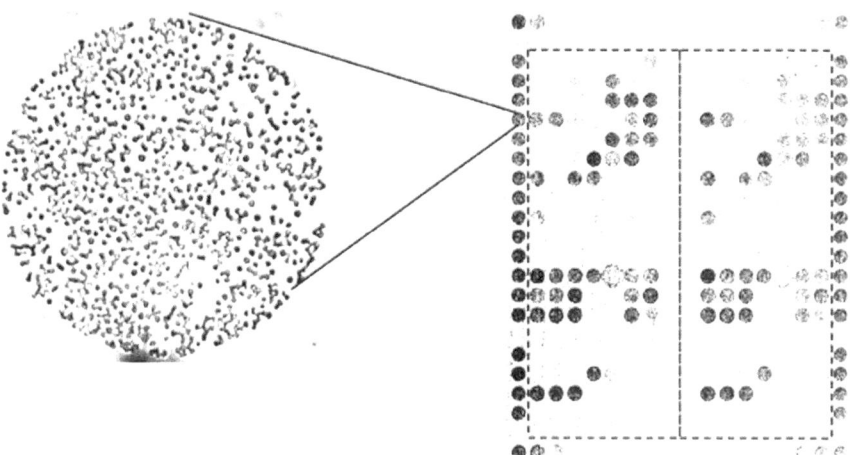

Fig. 3.1 Dark-field microscopy of a B cell sample analyzed on an antibody microarray. Anti-CD44 spots applied down the left and right sides of the duplicate arrays provide array alignment for defined antibody addresses. Anti-CD44 titrations are shown at the top and bottom of the array. The magnified image shows individual cells captured on a dot. Reprinted from *Cancer Letters,* 265, Ellmark P, Hogerkorp CM, Ek S, Belov L, Berglund M, Rosenquist R, Christopherson RI, and Borrebaeck CA, Phenotypic protein profiling of different B cell sub-populations using antibody CD-microarrays, 98–106, Copyright (2008), with permission from Elsevier

82 different antibodies on a chip (DotScan™ antibody microarray) and incubated T cells collected from human tonsils. After incubation and washing, the slides were scanned and images were acquired with an 8-bit camera (Fig. 3.1). Greyness levels for each spot were processed according to an ANOVA test and four clusters were identified and associated to B-lymphocyte subpopulations. These phenotypic profiles allowed the characterization of naïve B-lymphocytes, Germinal Center founder cells, Germinal Center cells, and memory B cells. mRNA expression levels of the targeted CD were also analyzed on genechips. DNA microarray data were similarly processed by running an ANOVA test, and in concordance with results obtained on antibody microarrays, 4 clusters corresponding to the same B cell subpopulations were distinguished. Among the 82 antibodies screened in this assay, 13 were clearly identified as key antigens for sorting the B cell subfamilies by antibody microarray analysis, 23 were identified as key antigens for sorting the B cells by gene expression analysis, and only 8 arose from both analyses. This feature thus strengthens the interest of using antibody microarrays for getting better insights in phenotypic analysis.

3.2.2 Typing Bacteria

CDs are often used as a target of choice for eukaryotic cell binding. But other surface antigens might also be selected in the search for contaminating bacteria, for instance. Due to its large social impact on public health and cost for society

(U.S. Department of Agriculture [29]), foodborne pathogens constitute a major issue in the development of fast, reliable, affordable, and easy-to-operate biosensors. Here, we will focus on the development of antibody microarrays for immunoassays developed to that end.

Since control assays are done on a routine basis, food-specific assays should allow highly parallelized analysis. This has been done, for instance, by microarraying antibodies in 96-multiwell microtiter plates [30]. After bacterial enrichment and incubation in the plates, *E. coli* and *Salmonella typhimurium* bacteria were successfully detected with 10^7 and 10^8 colony forming units (CFU)/mL. After labeling of the captured cells with a fluorophore-conjugated secondary antibody, immunoassays were automatically analyzed with a microplate reader. Distinguishing different bacterial strains or species is important, but sometimes is not sufficient: one of the most important examples is the characterization of *E. coli* strains, as some of them are harmless while other strains—Shiga Toxin-producing *E. coli* (STEC)—are the most researched foodborne pathogens. Among the STEC, six serogroups are frequently encountered and named "the big 6" [31, 32]. One must add the O157:H7 *E. coli* serogroup, which is also regularly involved in foodborne diseases. Recently Gehring et al. designed a high-throughput microarray capable of distinguishing these serotypes [32]. This antibody microarray was capable of specifically capturing each serotype—O26, O45, O103, O111, O121, O145, O157—with minimum cross-reactivity between each strain. The microarray readout was done by fluorescent scanning of the plates after bacteria capture and intracellular labeling by a nucleic acid targeting reporter molecule. However, bacterial serotyping on protein microarrays is not limited to STEC foodborne pathogens and can be extended to clinical applications. *Streptococcus pneumoniae* is responsible for thousands of deaths worldwide each year. One of the most important factors of pathogenicity of this bacterial strain is due to the composition of the capsule covering the bacterial surface. This surrounding layer is made of polysaccharides whose composition fires a specific immune response by the infected host. So far, 91 different *S. pneumoniae* serotypes have been identified [33]. Clustering their capsular antigenic determinants identifies these serogroups. In order to facilitate the easy determination of *S. pneumoniae* serotypes, Marimon et al. designed a protein microarray (Fig. 3.2) containing antibodies targeting 83 of the 91 serotypes described [34]. 226 mucoid isolates were analyzed (106 invasive isolates and a subset of 120 randomly selected noninvasive isolates) and data were compared with the results obtained by PCR analysis. Among this set of clinical samples, 37 different serotypes were distinguished among the 83 different serotypes targeted by the micro-arrayed antibodies and no statistically relevant differences were observed by comparison with standard methods used for *S. pneumoniae* serotyping.

Although most antibody microarrays used for bacteria typing or serotyping involve surface scanning for fluorescent signal analysis, other approaches have been described for microarray readout. For instance, surface-bound bacteria have been labeled by gold nanoparticles (GNP) functionalized by lectins [35]. By constructing a sandwich assay involving on one side an antibody/bacterial antigen recognition and a lectin/bacterial sugar binding on the other side, a high selectivity was obtained

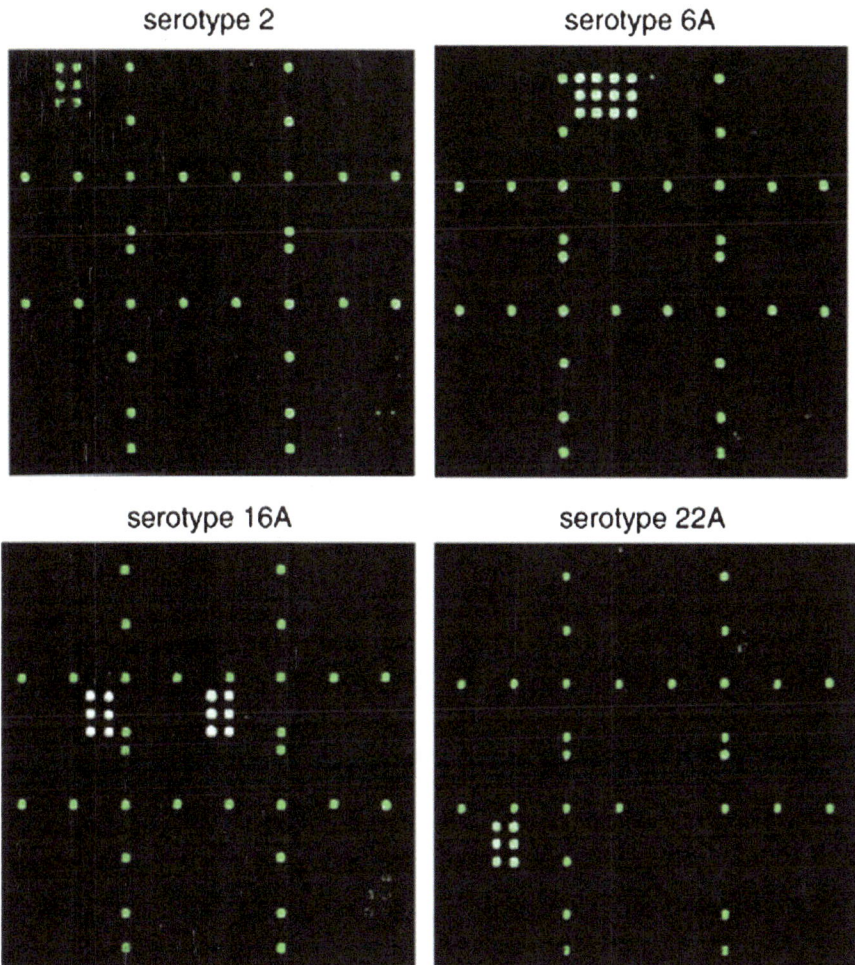

Fig. 3.2 *Streptococcus pneumonia* typing on antibody microarrays. After binding of fluorophore-labeled bacteria on microarray-wells, fluorescent signal is detected wherever the printed antibodies (specific to serotypes 2, 6A, 16A, and 22A) interact with the capsular antigens carried by the isolate being tested. The reference grid, made up of spots containing the fluorophore streptavidin-Cy3, allows easy identification of the antisera. Reprinted from *J Microbiol Methods* 80, Marimon JM, Monasterio A, Ercibengoa M, Pascual J, Prieto I, Simon L, and Perez-Trallero E, Antibody microarray typing, a novel technique for Streptococcus pneumoniae serotyping. 274–80, Copyright (2010), with permission from Elsevier

for *E. coli* and *Bacillus subtilis* strain detection. The labeling of bacteria with GNP followed by incubation with silver nitrate and hydroquinone initiates the electro-less silver deposition on the GNP and enables the identification of positive features by resonant light scattering (RLS). Such amplification processes gave significant results, as the limit of detection (LOD) was as low as 10^5 bacteria/mL. Although most antibody microarrays are analyzed after targeted labeling, there is a growing interest for label-free technologies, particularly on surface plasmon resonance (SPR) detection [26, 36–38]. Since samples are analyzed by the time they are deposited onto the microarray, real-time analysis of cell capture can be recorded. In the specific case of bacterial detection on microarray red by SPR, the usual LOD cannot be below the 10^4–10^5 CFU/mL range. Although such levels might be observed in naturally contaminated samples, international standards request the detection of contaminating bacteria present at concentrations as low as only a few bacteria per milliliter. The difference of several orders of magnitude between the desired concentration for detection in food and the LOD observed in standard bacterial assays requires one to first run an amplification step to enrich the sample with bacteria. This phase usually lasts for one night to a few days and thus significantly impacts the overall assay duration. Recently, Bouguelia et al. proposed an original approach based on a culture-capture-measure (CCM) of living bacteria on antibody microarrays [39]. The method is based on the incubation of raw sample—diluted to 1/10th in liquid culture medium according to standard methods—on a microarray and incubation at optimal temperature in an SPR imager. By monitoring in real-time the bacterial growth on the protein microarrays, remarkable LOD were reached with the detection of only a few Salmonella in less than 15 h. Interestingly, data analysis allowed the access to initial contamination levels, regardless of the strain detected in the sample. Due to the significant improvement in assay duration and easy operation, such label-free methodologies appear as promising approaches for living cell detection.

Besides food safety and clinical applications, the detection and characterization of bacteria in the environment might also be of significant interest. To that end, antibody microarrays have also been shown to be interesting tools, as they may appear as complementary to genetic-based assays (mostly PCR-based detection) [40, 41]. Series of soil samples (sediments, rock, granite) and water-based samples (ice, hydrothermal spring, mud) worldwide (Yellowstone National Park, Argentina desert, South African mine, Iceland, Antarctica, Atacama Desert, etc.) have been processed on antibody microarrays for detecting environmental bacteria. Using a 200-antibody microarray, Rivas et al. managed to map microbial biomarkers specific to the geological environment [41]. *Vibrio* spp., *Pseudomonas* spp., and *Actinobacteria* spp. were successfully identified and confirmed by control sample cloning and sequencing of the bacterial 16S rRNA gene. Although assays are undertaken on lysed bacteria—which supposedly improves the LOD—the bacterial concentration required for specific detection is still quite high (around 10^5 CFU/mL) by comparison to concentrations that might be encountered in their native environment. This limit thus implies the necessity of either a pre-concentration step or an enrichment step before running the microarray-based immunoassay. Nevertheless, the antibody microarray approach turns out to be an interesting tool for environmental applications, mainly due to the optimized storage protocols, and to the robustness of these microarrays, making them ideal candidates for extreme conditions such as planetary exploration [42, 43]. Dedicated instruments containing a sample preparation unit (for extraction and ultra-sonication) and a sample analysis unit (for fluorescent detection) were developed and optimized to detect bacteria and spores at levels of 10^3 particles/mL.

3.3   Characterizing Diseases Using Antibody Microarrays

Many pathologies trigger protein expression disorders. This is particularly the case with cancers, since they underlie several genetic disorders. By targeting specific cell surface markers, antibody microarrays turned out to be powerful tools for studying whole cells drawn from patients. From a practical point of view, fluorescence-assisted cell sorting (FACS) remains the technique of choice for probing the expression of cell surface markers. But, due to the limitation of excitation and emission spectra of fluorophores used in routine in flow cytometry, the multiplexed analysis of CD expression is often limited to 3–5 targets per cell sample. On the other hand, thanks to the high-throughput arraying of antibodies on surfaces, protein microarray developments are more often limited by the availability of a large repertoire of CD-specific antibodies than by surface density saturation.

The antibody microarray approach for disease or pathological disorder monitoring has been developed for more than a decade by Belov et al. on leukocytes from healthy or sick patients [44]. Early studies have been carried out on a series of 60 antibodies micro-arrayed on nitrocellulose membranes incubated with leukocytes. The fluorescent labeling of captured cells with either anti-CD3 FITC or anti-CD19 PE antibodies allowed the identification of T and B lymphocytes, respectively. Authors demonstrated that several stages of leukemia disease could be diagnosed by comparing signal patterns with responses from normal pre-sorted leukocytes. The results were confirmed by analysis on cDNA microarrays and flow cytometry analysis. Although protein microarrays appear to be a powerful tool for whole-cell analysis, one should indicate that this technology is not intended to replace flow cytometry, particularly when several questions have to be addressed at the individual cell level; as cells are captured on features functionalized by a single probe, the direct capture of cells only informs the presence of one CD on its surface. Sandwich assays—with the limitation of emission/excitation spectra—with fluorescently labeled antibodies might be done but cannot be extended to a large number of targets as signal deconvolution requires large computing resources. Similar experiments were also led on larger antibody collections (82 micro-arrayed antibodies) and allowed the identification of normal B cell populations and malignant B cell populations [28]: after cell binding, the specific response of each feature allowed a principle component analysis that clearly identified germinal center founder cell populations and normal naïve B cells. More recently, this group published results obtained on antibody microarrays with higher number of probes (147 antibodies) still targeting CD antigens. This new generation of microarrays (named DotScan™ microarrays) allowed the identification of expression profiles specific to discrete leukemia subtypes [45]. The thorough analysis of the collected data outlined several differences between normal leukocytes and acute myeloid leukemia (AML) with AMC cells expressing the myeloid antigens (CD13, CD15, and CD33), the stem cell precursor markers (CD43 and CD117) and a range of other antigens (e.g., CD4, which is usually highly specific to T lymphocytes). Once again, the authors outline the need to run counter labeling with appropriate antibodies to strengthen the data issued from one subpopulation.

Leukocyte populations can also be interrogated for non-cancer disorders; recently, an 82-antibody microarray has been designed and tested for the diagnosis of systemic lupus erythematosus (SLE) [46]. SLE is an immune inflammatory disease characterized by abnormal autoantibody production. This disorder is often difficult to characterize as the diagnosis is based on clinical symptoms that may take years to appear. Soluble biomarkers have been identified (serum complement C3 and C4 and anti-dsDNA antibody) but they are not sufficient to reliably characterize this disease. Using a specific set of immobilized antibodies, PBMCs from both healthy donors and SLE-diagnosed patients, these were incubated on protein microarrays. Interestingly, the data analysis not only allowed one to distinguish these two patient populations, but it also allowed the separation of patients with SLE from patients with other autoimmune disorders.

As explained in the introduction of this chapter, most cellular assays done on antibody microarrays are done with circulating cells. However, adherent cells have also been processed on protein microarrays for disease diagnosis [47, 48]. Kaufman et al. prepared a microarray specific for metastatic melanoma and processed clinical lymph node samples [47]. Such samples contain a large variety of cell types including cancer cells, leukocytes, and also lymph node stromal cells and endothelial cells. To increase the cell subpopulation specificity, samples were treated with anti-CD45 antibody-coated magnetic beads that allowed melanoma cells and leukocytes to be enriched and profiled separately. Other adherent cells have been studied on protein microarrays for determining disease-specific expression profiles. Liu-spotted antibodies (119 CD-specific antibodies) on polystyrene Petri dishes and screened several prostate cancer cell lines [49]. Although some heterogeneity was observed in expression profiles, abnormal profiles specific to each cancer cell line were identified.

Besides diagnosis applications, there has been a step forward in the monitoring of disease stages, such as in the case of HIV patients [50]. CD4- and CD8-expressing lymphocytes are commonly used as markers for HIV infection evolution in patients. To that end, absolute and relative counting of these T-lymphocyte populations is regularly followed. Belov's group studied purified CD4- and CD8-expressing sets of cells upon incubation on antibody microarrays. Three pools of blood samples were chosen: one from long-term non-progressor patients naturally controlling viremia; one from HIV+ patients under highly active antiretroviral therapy (HAART) with low viral load (<50 copies/mL); and the last from HIV patients under HAART with viremia. Experiments were led on microarrays functionalized with 135 antibodies targeting surface antigens. Pairwise analysis of the results allowed the identification of an antigen set containing 17 antigens, 5 antigens of which have never been described so far as involved in HIV disease progression. One year later, this group published other results of interest on the application of such microarrays for HIV-infected patient follow-up [51]. In this manuscript, there were two pools of HIV patients with HAART: one group of patients achieving sustainable response by maintaining below detectable plasma viremia, and the other responding intermittently to the treatment. Among the 135 antibodies used on this device, the expression of seven markers was common to the combined groups, whereas the expression levels of ten markers were different for these pools. Interestingly, expression levels of CD9, CD11a, CD27, CD28, and CD11c, CD44, CD49d, CD49e, and CD54 were directly linked to CD4 and CD8 cell counts, respectively.

3.4   Functional Antibody Microarrays: Application to Cell Secretion Detection

3.4.1   Analyzing Cellular Populations on a Protein Microarray

As described earlier in this chapter, protein microarrays are divided into two major classes: analytical and functional devices [52]. The interest in developing functional microarrays is partly due to the benefits of assay miniaturization; indeed, by decreasing sample volumes, the surface-to-volume ratio increases dramatically. This specific feature thus motivated surface protein immobilization processes to go one step further into cellular biological insights [53]. By targeting circulating blood cells, biologists and physicians hope to get more details on physiological events occurring at the cellular level. Since flowing cells have poor physical interference with other cells, one of the most important physiological phenomena is the production and secretion of soluble products that react with other cells at a later time, and in this manner channel information through the body. The library of secreted products—sometimes referred to as a secretome—includes mainly hormones and cytokines. Hormones remain an important target of secreted products in many diagnosis assays. Their main characteristics reside in the fact that hormones are secreted by a specific cell type and are diffused through the body before reaching one unique target cell type. In the case of immunological responses, which are also of primary interest [54–56], the problem at hand shows a much higher degree of complexity, as the same cytokine might be secreted by several cell types and may target different cell types. The large repertoire of both cytokines and biological events fired by these soluble factors explains their importance for clinical diagnosis [54]. ELISAs are very popular assays because they allow the sensitive and qualitative detection of cytokines in vitro, although no information on secreting cells is produced. In vitro culture of immune cells and measurements of their reactivity toward extracellular agents (antigenic stimulation for instance) at the cellular level is also of important interest [57, 58]. This is motivated by the evidence that in vitro measurements of cytokine levels might be directly correlated to in vivo lymphocyte activity [59].

Besides whole-cell recognition and phenotype characterization, antibody microarrays have also been used in functional cell-based assays and more specifically, for the monitoring of cytokine release in the extracellular medium [60]. To that end, Kasai et al. developed glass slides functionalized with anti-IL1b antibody and deposited islets of cells entrapped in collagen gel droplets [61]. This approach permits the deposition of several cell types on a single microarray and allows interrogation of them for the production of one cytokine. The plotting of a calibration curve obtained by incubating known amounts of cytokine and scanning electrochemical microscopy (SECM) analysis allowed the quantitative assessment of cytokine production at the droplet level. However, cell-specific capture and cytokine detection might also be done on the same antibody microarray [27, 62]. Revzins' group proposed biochips functionalized with series of spots containing antilymphocytes antibodies (CD4 and CD8) and spotted alongside antibodies specific to IFNγ and IL-2 cytokines [27]. After T lymphocyte capture on the spots and incubation, cytokines spontaneously diffused and were immobilized on cytokine-specific surrounding spots. Both CD4+ and CD8+ subpopulations were independently studied for each cytokine revealed in ELISA-like sandwich labeling. The same approach was also validated on hepatocytes (adherent cells), processed in a similar manner and interrogated for α1-antitrypsin production [62]. In this latter example, hepatocytes were immobilized using a protein from the extracellular matrix (collagen) rather than with an antibody targeting cell surface antigens. In an attempt to describe original antibody-based multiplexed platforms, some authors also engineered DNA-conjugated antibodies for microarray applications [36, 63]. This strategy was successfully applied to the simultaneous capture of cells and detection of secreted products [63] and showed remarkable LOD (10 fM for IL-2) due to a nanoparticle amplification effect.

Functional assays dedicated to immune responses after antigenic stimulation are important for control of vaccination efficiency, and also for epitope screening or identification of multiple epitope-specific T cell populations. An elegant way to sort these cellular subpopulations is based on the immobilization on a surface of peptides bound to class I or class II MHC (major histocompatibility complex) molecules rather than antibodies targeting CD [64]. Such an approach showed significant results, thus validating the use of such protein arrays for the identification of rare T lymphocytes from vaccinated mice. More recently, the same group increased the complexity of such microarrays by combining cytokine-specific antibodies to peptide-MHC [65]. Blood drawn from patients was processed on a biochip, captured cells were stimulated upon pMHC binding, and T cell responses were monitored (secretion of IFNγ, TNFα, Granzyme B, IP-10, IL-2, IL6, and IL-1b). Samples from patients suffering from melanoma and who received vaccination showed statistically different profiles than those who did not receive vaccination. One interesting conclusion of this work is the correlation between high levels of IFNγ and TNFα and the arrest of tumor development. Such tools thus provide an easy-to-process device for the fast assessment of vaccination efficiency.

A common feature for the example of functional protein microarrays presented here is the requirement of a labeling step for cytokine production read-out. Although fluorescent labeling or nanoparticle conjugation often permits an improvement to the LOD, the end-point analysis impedes access to kinetic parameters. Due to the very low amounts of cytokines secreted by few hundreds of cells immobilized on a surface, there are only a few examples of label-free detection of cytokines on protein microarrays [38, 66]. SPR turns out to be a suitable approach for monitoring cytokine capture on antibody microarrays; either when the analysis is done downstream of the cell-specific capture [66] or when cell and cytokine capture are completed on a single biochip [38].

3.4.2   On the Use of Microtechnologies for Designing New Functional Microarrays

Since the last two decades, device miniaturization for biomedical applications benefited from the developments in micro- and nanotechnologies, in particular for surface functionalization. For instance, micro-contact printing using polymeric molds enabled the deposition of proteins on features at the micrometer level [67]. Interestingly, by patterning a surface at length scales smaller than individual cells, new insights became accessible. This can be illustrated by the work of Shen et al. who patterned surfaces with different proteins and followed the secreting profiles as a function of stimulating protein localization (Fig. 3.3). During immune cascades involving T cells, it is known that cell–cell communication occurs mainly through small intercellular contacts (>100 μm^2) termed immunological synapses (IS). These synapses involve protein complexes, which will interact in a given order to eventually activate the T cell. Molecular studies identified the involvement of several complexes, specifically the TCR and CD28 complexes. Cellular co-stimulation, often initiated by complex binding, seems to be more important when CD28 complexes are segregated and surrounding the IS. This hypothesis was tested by arraying central features with anti-CD3 antibodies (binding and activating the TCR complex), and microarraying smaller features—functionalized with anti-CD28 antibodies—in an intermingled matrix. Monitoring IL-2 production at the individual cell level assessed stimulation efficiency. Data analysis confirmed that presenting anti-CD28 antibodies in the cell periphery of anti-CD3 spots enhances the IL-2 secretion by CD4+ T lymphocytes. These conclusions were strengthened by control experiments led on similar arrays where each feature contained a mixture of both antibodies, with the result being that they did not trigger a significant secreting amplification effect.

3.4.3   Analyzing Individual Cellular Responses on an Antibody Microarray

Another significant advance in cell-based assays due to recent improvements in microtechnology is the access to individual cell responses. On the contrary to single-cell assays where one unique object is observed [68], individual-cell assays enable the monitoring of physiological events occurring at the single-cell level from a large number of individual cells. The access to individual responses measured for each member of a cell population is important from a biological point of view, as significant variations have been observed even through genetically similar cells [69]. Surface patterning techniques permitted the creation of large series of micro-chambers where individual cells can be isolated and cultured for a few hours [70, 71]. Han et al. designed and fabricated parallelized micro-cuvettes by micro-engraving and incubating individual immune cells on this array. After covering the array with a cover glass slide functionalized with four different antibodies

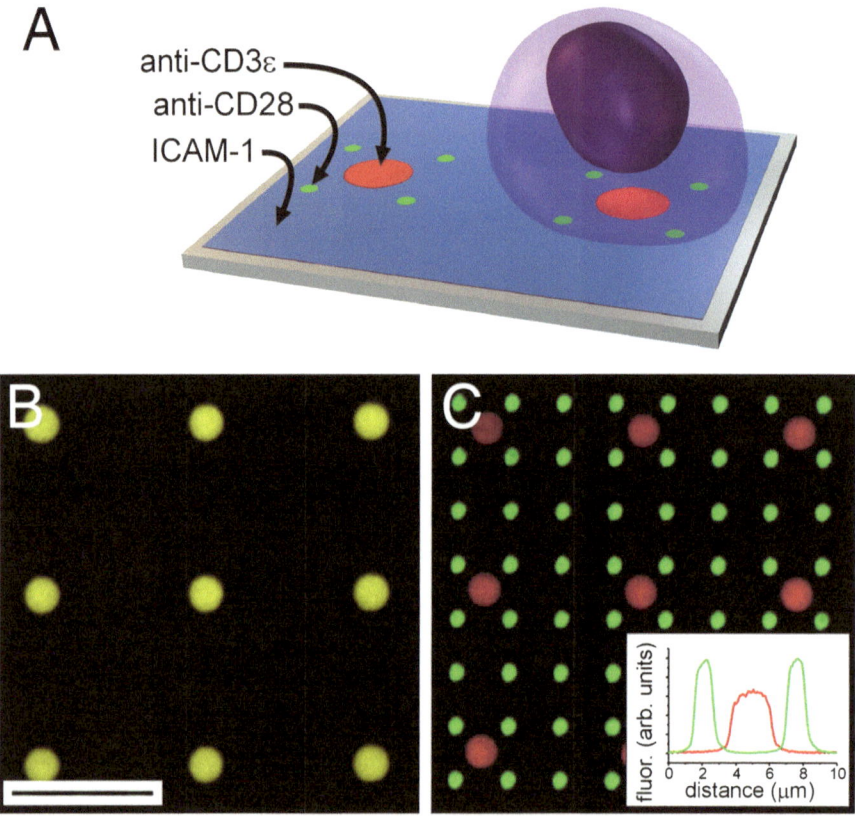

Fig. 3.3 Multiplexed detection of T cell interaction with micro-patterned, co-stimulatory arrays. (**a**) CD4$^+$ cells were presented with micro-arrayed antibodies that capture the microscale organization of ligands associated with T cell co-stimulation. Colocalized patterns were created by mixing anti-CD3 and anti-CD28 antibodies (*yellow*) in a single step (**b**), while segregated patterns were defined by sequential patterning of anti-CD3 (*red*) and anti-CD28 (*green*) on a single surface (**c**). (*Inset*) Fluorescence profile across a segregated site. ICAM-1 was coated onto the remainder of these surfaces but is omitted here for clarity (scale bar: 10 μm). Reprinted from Shen K, Thomas VK, Dustin ML, and Kam LC (2008) Micropatterning of costimulatory ligands enhances CD4+ T cell function. *Proc Natl Acad Sci U S A* 105, 7791–6 with permission from PNAS

(anti-IL-6, anti-IL-17, anti-IFNγ, anti-IL-2, and anti-TNFα), secretion levels were determined by molecular sandwich-like construction with fluorescently labeled antibodies and secreting rates per cell were calculated for each cytokine. Microfluidics also significantly impacted individual cell-based assays by enabling the channeling and isolation of individual cells [72] (Situma Biomol 2006). On an experimental point of view, the usage of polydimethylsiloxane (PDMS) molds paved the way to a large panel of applications focused on cell analysis. This is illustrated either by static assays where microstructures (Fig. 3.4) were easily processed by PDMS molding [71] or on dynamic assays where microstructures are used for injecting, separating, and studying cells on antibody microarrays [73, 74]. In their former work, Heaths'

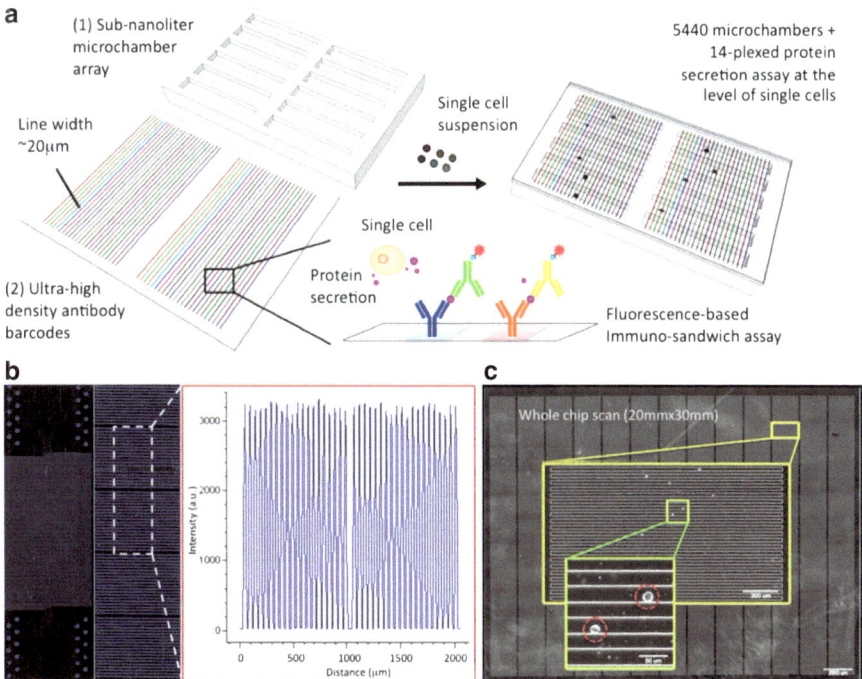

Fig. 3.4 A high-throughput multiplexed single cell secretomic assay on antibody microarray. (**a**) Schematic illustration showing integration of a high-density antibody barcode array chip and a subnanoliter micro-chamber array chip for a protein secretion assay at the single cell level. (**b**) Scanned fluorescence image showing high uniformity of protein loading across the entire barcode microarray. (**c**) Photograph stitched from a large number of individual pictures collected by an automated, motorized phase contrast microscope. Scale bar 2 mm. The first enlarged image shows a column of micro-chamber array (scale bar 300 µm). The second enlarged image shows individual cells loaded in micro-chambers (scale bar 50 µm). Reprinted with permission from Lu Y, et al. *Analytical Chemistry,* (2013), 85(4):2548–2556. Copyright (2013) American Chemical Society

group fabricated a microfluidic platform designed for multiplexed analysis of individual cells isolated in more than 1,000 nL-volume micro-chambers [73]. This device is composed of 80 microchannels actuated by valves, enabling the isolation of more than 10^4 cells in individual reaction chambers. Each chamber is functionalized by 12 stripes of proteins specific to secreted products. Human macrophages were chemically activated for cytokine production and production of IFNγ, TNFα, IL-2, IL-1a, IL-1b, IL-6, IL-10, and IL-12 was subsequently measured. Control experiments were led by intra-cellular cytokine staining (ICS) of cytokines and flow cytometry. This analysis emphasized the discrepancy of secreting and non-secreting cells, although macrophages were differentiated from the THP-1 human cell line. A similar experiment was also undertaken with PBMC cells, enriched for CD3 and CD8 markers. The authors observed a high level of functional heterogeneity within this subpopulation sorted with these two markers. The simultaneous assessment of secreting properties for the set of twelve proteins confirmed the polyfunctional property of these cells, since an average of more than five different secreted proteins were detected. Such a complex network of secretion profiles fuels the need for new tools suitable for the multiplexed analysis of individual cells.

3.5   Conclusion

The development of surfaces functionalized with series of probes in the 1980s paved the way for a tremendous number of applications based on biomolecular recognition. The parallelized analysis of a large number of assays, conjugated to recent developments in microtechnologies, i.e., microfluidics, brought a large number of advantages such as reduction in regent cost, enhancement in assay speed, high potential for mass production of devices at low cost, the ability to integrate several processing steps into a single system, and easy operation by untrained technicians. Although early examples of microarrays were described with immobilized proteins for parallelized immune-recognition assays, the most rapid development in the biomedical field was observed with DNA microarrays. This feature might be partly explained by some limitations linked to protein (mostly antibody) handling, which are more expensive to produce than short DNA strands, more difficult to site-specifically immobilize on a surface, and more sensitive to environmental conditions (buffer, salt concentration, humidity, adjuvants, etc.) after deposition. However, proteins are involved in almost all biochemical processes that occur in a cell, most of which cannot be investigated by utilizing DNA microarrays. DNA biochips thus fail to shed light on events occurring downstream of protein expressions. Significant efforts have been made in engineering alternative probe molecules exhibiting specific binding properties and chemical stability, along with easy chemical derivatization. Among potential ligands, aptamers have had much focused attention and effort for more than two decades, but so far have failed in replacing proteins in biochemical assays used on a routine basis in analysis laboratories [75]. The unique binding properties and diversity of proteins explain the constant development of protein microarrays [76]. Among the proteins used as probes immobilized on surfaces, antibodies represent an abundant class of molecules deposited on microarrays. As demonstrated in the very first example of microarrays [15], the targeting of whole native cells is an important issue in several biomedical applications. More sophisticated antibody microarrays are arising, pushed forward by the developments led in parallel in micro- and nanotechnologies [72]. The recent developments remain focused on whole eukaryotic or prokaryotic cell identification, but may also allow access to physiological responses, as illustrated in cellular detection assays carried out on antibody biochips. However, cell-based miniaturized assays do not allow the investigation regarding physiological events at the individual cell level. An emblematic example is the search for rare cells—either stem cells or CTCs—that remain highly challenging to detect on microarrays because they are present at low levels in a biological sample and only a tiny amount of material is usually loaded on each run.

Nevertheless, antibody microarrays continue to raise interest for simple, robust, fast, and specific assays, as illustrated by recent examples of biochip fabrication for space missions and extraterrestrial life detection on Mars [43].

References

1. Jin C et al (2014) Technologies for label-free separation of circulating tumor cells: from historical foundations to recent developments. Lab Chip 14(1):32–44
2. Haeberle S, Zengerle R (2007) Microfluidic platforms for lab-on-a-chip applications. Lab Chip 7(9):1094–1110
3. Grant SF, Hakonarson H (2008) Microarray technology and applications in the arena of genome-wide association. Clin Chem 54(7):1116–1124
4. Zhu H, Qian J (2012) Applications of functional protein microarrays in basic and clinical research. Adv Genet 79:123–155
5. Foong YM, Fu J, Yao SQ, Uttamchandani M (2012) Current advances in peptide and small molecule microarray technologies. Curr Opin Chem Biol 16(1–2):234–242
6. Oyelaran O, Gildersleeve JC (2009) Glycan arrays: recent advances and future challenges. Curr Opin Chem Biol 13(4):406–413
7. Schena M, Shalon D, Davis RW, Brown PO (1995) Quantitative monitoring of gene expression patterns with a complementary DNA microarray. Science 270(5235):467–470
8. Fodor SP et al (1991) Light-directed, spatially addressable parallel chemical synthesis. Science 251(4995):767–773
9. Ekins R, Chu F, Biggart E (1989) Development of microspot multi-analytite ratiometric immunoassay using dual fluorescent-labelled antibodies. Anal Chim Acta 227:73–96
10. Weller MG (2003) Classification of protein microarrays and related techniques. Anal Bioanal Chem 375(1):15–17
11. Ekins R, Chu F, Biggart E (1990) Fluorescence spectroscopy and its application to a new generation of high sensitivity, multi-microspot, multianalyte, immunoassay. Clin Chim Acta 194(1):91–114
12. Kodadek T (2001) Protein microarrays: prospects and problems. Chem Biol 8(2):105–115
13. Lynes MA (2005) Solid-phase immunoassays. Curr Protoc Toxicol Chapter 18:Unit18 17. Wiley
14. Dexlin-Mellby L et al (2011) Design of recombinant antibody microarrays for membrane protein profiling of cell lysates and tissue extracts. Proteomics 11(8):1550–1554
15. Chang TW (1983) Binding of cells to matrixes of distinct antibodies coated on solid surface. J Immunol Methods 65(1–2):217–223
16. Castel D, Pitaval A, Debily MA, Gidrol X (2006) Cell microarrays in drug discovery. Drug Discov Today 11(13–14):616–622
17. Chen DS, Davis MM (2006) Molecular and functional analysis using live cell microarrays. Curr Opin Chem Biol 10(1):28–34
18. El-Ali J, Sorger PK, Jensen KF (2006) Cells on chips. Nature 442(7101):403–411
19. Utz PJ (2005) Protein arrays for studying blood cells and their secreted products. Immunol Rev 204:264–282
20. Kato K, Toda M, Iwata H (2007) Antibody arrays for quantitative immunophenotyping. Biomaterials 28(6):1289–1297
21. Flaim CJ, Chien S, Bhatia SN (2005) An extracellular matrix microarray for probing cellular differentiation. Nat Methods 2(2):119–125
22. Kim H, Doh J, Irvine DJ, Cohen RE, Hammond PT (2004) Large area two-dimensional B cell arrays for sensing and cell-sorting applications. Biomacromolecules 5(3):822–827

23. Revzin A, Sekine K, Sin A, Tompkins RG, Toner M (2005) Development of a microfabricated cytometry platform for characterization and sorting of individual leukocytes. Lab Chip 5(1):30–37
24. Ko IK, Kato K, Iwata H (2005) Antibody microarray for correlating cell phenotype with surface marker. Biomaterials 26(6):687–696
25. Fujii Y, Anderson JM, Matsuda T (2008) Antibody-bound cell microarray for immunophenotyping: surface modification and lymphocyte subpopulations. J Biomed Mater Res B Appl Biomater 87(2):525–537
26. Suraniti E et al (2007) Real-time detection of lymphocytes binding on an antibody chip using SPR imaging. Lab Chip 7(9):1206–1208
27. Zhu H et al (2008) A microdevice for multiplexed detection of T-cell-secreted cytokines. Lab Chip 8(12):2197–2205
28. Ellmark P et al (2008) Phenotypic protein profiling of different B cell sub-populations using antibody CD-microarrays. Cancer Lett 265(1):98–106
29. Anonymous (2010) Response to questions posed by the food safety and inspection service regarding determination of the most appropriate technologies for the food safety and inspection service to adopt in performing routine and baseline microbiological analyses. J Food Prot 73(6):1160–1200
30. Gehring AG, Albin DM, Reed SA, Tu SI, Brewster JD (2008) An antibody microarray, in multiwell plate format, for multiplex screening of foodborne pathogenic bacteria and biomolecules. Anal Bioanal Chem 391(2):497–506
31. Gehring A et al (2013) A high-throughput antibody-based microarray typing platform. Sensors (Basel) 13(5):5737–5748
32. Hegde NV, Praul C, Gehring A, Fratamico P, Debroy C (2013) Rapid O serogroup identification of the six clinically relevant Shiga toxin-producing Escherichia coli by antibody microarray. J Microbiol Methods 93(3):273–276
33. Park IH et al (2007) Discovery of a new capsular serotype (6C) within serogroup 6 of Streptococcus pneumoniae. J Clin Microbiol 45(4):1225–1233
34. Marimon JM et al (2010) Antibody microarray typing, a novel technique for Streptococcus pneumoniae serotyping. J Microbiol Methods 80(3):274–280
35. Gao J, Liu C, Liu D, Wang Z, Dong S (2010) Antibody microarray-based strategies for detection of bacteria by lectin-conjugated gold nanoparticle probes. Talanta 81(4–5):1816–1820
36. Bombera R, Leroy L, Livache T, Roupioz Y (2012) DNA-directed capture of primary cells from a complex mixture and controlled orthogonal release monitored by SPR imaging. Biosens Bioelectron 33(1):10–16
37. Dudak FC, Boyaci IH (2009) Rapid and label-free bacteria detection by surface plasmon resonance (SPR) biosensors. Biotechnol J 4(7):1003–1011
38. Milgram S et al (2011) On chip real time monitoring of B-cells hybridoma secretion of immunoglobulin. Biosens Bioelectron 26(5):2728–2732
39. Bouguelia S et al (2013) On-chip microbial culture for the specific detection of very low levels of bacteria. Lab Chip 13(20):4024–4032
40. Parro V et al (2011) A microbial oasis in the hypersaline Atacama subsurface discovered by a life detector chip: implications for the search for life on Mars. Astrobiology 11(10):969–996
41. Rivas LA et al (2008) A 200-antibody microarray biochip for environmental monitoring: searching for universal microbial biomarkers through immunoprofiling. Anal Chem 80(21):7970–7979
42. de Diego-Castilla G et al (2011) Assessing antibody microarrays for space missions: effect of long-term storage, gamma radiation, and temperature shifts on printed and fluorescently labeled antibodies. Astrobiology 11(8):759–773
43. Parro V et al (2011) SOLID3: a multiplex antibody microarray-based optical sensor instrument for in situ life detection in planetary exploration. Astrobiology 11(1):15–28
44. Belov L, de la Vega O, dos Remedios CG, Mulligan SP, Christopherson RI (2001) Immunophenotyping of leukemias using a cluster of differentiation antibody microarray. Cancer Res 61(11):4483–4489

45. Barber N et al (2009) Profiling CD antigens on leukaemias with an antibody microarray. FEBS Lett 583(11):1785–1791
46. Lin MW, Ho JW, Harrison LC, dos Remedios CG, Adelstein S (2013) An antibody-based leukocyte-capture microarray for the diagnosis of systemic lupus erythematosus. PLoS One 8(3):e58199
47. Kaufman KL et al (2010) An extended antibody microarray for surface profiling metastatic melanoma. J Immunol Methods 358(1–2):23–34
48. Rahman W et al (2012) Analysis of human liver disease using a cluster of differentiation (CD) antibody microarray. Liver Int 32(10):1527–1534
49. Liu AY (2000) Differential expression of cell surface molecules in prostate cancer cells. Cancer Res 60(13):3429–3434
50. Wu JQ et al (2007) Antibody microarray analysis of cell surface antigens on CD4+ and CD8+ T cells from HIV+individuals correlates with disease stages. Retrovirology 4:83
51. Wu JQ et al (2008) Longitudinal microarray analysis of cell surface antigens on peripheral blood mononuclear cells from HIV+individuals on highly active antiretroviral therapy. Retrovirology 5:24
52. Sutandy FX, Qian J, Chen CS, & Zhu H (2013) Overview of protein microarrays. Current protocols in protein science/editorial board, John E. Coligan ... [et al.] Chapter 27: Unit 27 21. Wiley
53. Wilson DS, Nock S (2002) Functional protein microarrays. Curr Opin Chem Biol 6(1): 81–85
54. Bienvenu J, Monneret G, Fabien N, Revillard JP (2000) The clinical usefulness of the measurement of cytokines. Clin Chem Lab Med 38(4):267–285
55. Hernandez-Fuentes MP, Warrens AN, Lechler RI (2003) Immunologic monitoring. Immunol Rev 196:247–264
56. Schett G, Elewaut D, McInnes IB, Dayer JM, Neurath MF (2013) How cytokine networks fuel inflammation: toward a cytokine-based disease taxonomy. Nat Med 19(7):822–824
57. Bromage E, Stephens R, Hassoun L (2009) The third dimension of ELISPOTs: quantifying antibody secretion from individual plasma cells. J Immunol Methods 346(1–2):75–79
58. Cox JH, Ferrari G, Janetzki S (2006) Measurement of cytokine release at the single cell level using the ELISPOT assay. Methods 38(4):274–282
59. Brattig NW et al (2005) Relevance of ex vivo blood lymphocyte assay for in vivo lymphocyte function. Clin Exp Immunol 139(1):127–131
60. Gonzalez LC (2012) Protein microarrays, biosensors, and cell-based methods for secretome-wide extracellular protein-protein interaction mapping. Methods 57(4):448–458
61. Kasai S et al (2005) Real-time monitoring of reactive oxygen species production during differentiation of human monocytic cell lines (THP-1). Anal Chim Acta 549(1–2):14–19
62. Jones CN et al (2008) Multifunctional protein microarrays for cultivation of cells and immunodetection of secreted cellular products. Anal Chem 80(16):6351–6357
63. Bailey RC, Kwong GA, Radu CG, Witte ON, Heath JR (2007) DNA-encoded antibody libraries: a unified platform for multiplexed cell sorting and detection of genes and proteins. J Am Chem Soc 129(7):1959–1967
64. Soen Y, Chen DS, Kraft DL, Davis MM, Brown PO (2003) Detection and characterization of cellular immune responses using peptide-MHC microarrays. PLoS Biol 1(3):E65
65. Chen DS et al (2005) Marked differences in human melanoma antigen-specific T cell responsiveness after vaccination using a functional microarray. PLoS Med 2(10):e265
66. Stybayeva G, Kairova M, Ramanculov E, Simonian AL, Revzin A (2010) Detecting interferon-gamma release from human CD4 T-cells using surface plasmon resonance. Colloids Surf B Biointerfaces 80(2):251–255
67. Shen K, Thomas VK, Dustin ML, Kam LC (2008) Micropatterning of costimulatory ligands enhances CD4+ T cell function. Proc Natl Acad Sci U S A 105(22):7791–7796
68. Andersson H, van den Berg A (2004) Microtechnologies and nanotechnologies for single-cell analysis. Curr Opin Biotechnol 15(1):44–49

69. Niepel M, Spencer SL, Sorger PK (2009) Non-genetic cell-to-cell variability and the consequences for pharmacology. Curr Opin Chem Biol 13(5–6):556–561
70. Han Q, Bradshaw EM, Nilsson B, Hafler DA, Love JC (2010) Multidimensional analysis of the frequencies and rates of cytokine secretion from single cells by quantitative microengraving. Lab Chip 10(11):1391–1400
71. Lu Y et al (2013) High-throughput secretomic analysis of single cells to assess functional cellular heterogeneity. Anal Chem 85(4):2548–2556
72. Situma C, Hashimoto M, Soper SA (2006) Merging microfluidics with microarray-based bioassays. Biomol Eng 23(5):213–231
73. Ma C et al (2011) A clinical microchip for evaluation of single immune cells reveals high functional heterogeneity in phenotypically similar T cells. Nat Med 17(6):738–743
74. Shi Q et al (2012) Single-cell proteomic chip for profiling intracellular signaling pathways in single tumor cells. Proc Natl Acad Sci U S A 109(2):419–424
75. Baird GS (2010) Where are all the aptamers? Am J Clin Pathol 134(4):529–531
76. Hall DA, Ptacek J, Snyder M (2007) Protein microarray technology. Mech Ageing Dev 128(1):161–167

Chapter 4
Polymer Microarrays for High Throughput Biomaterials Discovery

Andrew L. Hook

Abstract High throughput screening has emerged as a powerful technique for discovering novel medical tools and therapies. This is particularly true for biomaterials that are applied to poorly understood biological–material systems. The polymer microarray format has become a key enabling tool for high throughput materials discovery, whereby hundreds to thousands of unique polymers can be presented on a single glass slide and screened in parallel for biological interactions of interest. This approach has successfully been utilized to develop the surface chemistry, topography, bioactivity, and mechanical properties of biomaterials as well as allowing the development of 3D culture systems. In order to optimize a polymer microarray for a given application the substrate used, the coating on the substrate, and the material library screened must be carefully selected. Furthermore, development of suitable biological assays with high throughput readouts is imperative for expanding the applications of polymer microarrays. The biological systems screened on this format include supporting cell attachment and outgrowth, maturation and phagocytosis of dendritic cells, materials resistant to microbes, switchable materials, platelet activation, cell sorting, hepatocytes and toxicity models, and cell transfection. Further to the discovery and development of biomaterials, the large datasets when coupled with modelling techniques can establish structure–function relationships that help elucidate the underlying biological–material interactions. Continued development of microarray designs and high throughput biological assays compatible with the format will lead to the discovery of new biomaterials that exhibit unprecedented control over the biological systems they are designed to function in.

A.L. Hook (✉)
Laboratory of Biophysics and Surface Analysis, University of Nottingham,
Nottingham NG7 2RD, UK
e-mail: Andrew.Hook@nottingham.ac.uk

E. Palmer, *Cell-Based Microarrays: Review of Applications,*   53
Developments and Technological Advances, SpringerBriefs in Cell Biology,
DOI 10.1007/978-1-4939-0594-2_4, © Ella Palmer 2014

Abbreviations

2D/3D	2/3 Dimensions
AFM	Atomic force microscopy
BMDC	Bone marrow dendritic cells
ECM	Extracellular matrix
FBS	Fetal bovine serum
hEB	Human embryoid bodies
hES	Human embryonic stem cell
hMSC	Human mesenchymal stem cell
mES	Mouse embryonic stem cell
PEG	Poly(ethylene glycol)
pHEMA	Poly(hydroxyethyl methacrylate)
PLS	Partial least square
SEE	Standard error of estimation
SPR	Surface plasmon resonance
ToF-SIMS	Time-of-flight secondary ion mass spectrometry
WCA	Water contact angle
XPS	X-ray photoelectron spectroscopy

4.1 Introduction

Hypothesis led research has been a foundation of scientific endeavour since the seventeenth century. In some cases the level of understanding is not at a sufficient level to enable the formulation of pertinent hypotheses to effectively promote a field of science. In these cases a complementary data-driven approach has emerged, where the starting point is the accumulation of data that will, in turn, lead to new ideas [1, 2]. This approach has been applied to drug discovery projects that have used high throughput methods to screen for novel drugs [3–5]. High throughput screening has also been applied successfully to genomics [6–11], which has driven the emergence of the microarray format as a screening tool. On a microarray hundreds to thousands of unique analytes can be displayed on a sample slide in address-able locations and assessed in parallel. More recently, cell-based microarrays, where cells themselves are arrayed onto a slide or cells are allowed to interact with the arrayed analyte, have been developed to explore gene expression in systems where all the cell machinery is present to ensure correct protein function [12–16]. A num-ber of studies have applied microarrays to materials, with a focus on the discovery of novel materials that are ideally suited to a given application. It is this body of work that this chapter will cover.

4.2   The Development of a Material Microarray

The concept of a materials array was first reported in 1995 [17], whereupon an array of solid-state materials was produced by vacuum deposition onto areas selected by the use of a series of binary masks. A 128-member array was produced from seven precursor materials and two novel superconducting films, BiSrCaCuO and YBaCuO, were identified. Typical sample size was 2×2 mm, however, samples as small as 200×200 µm were also generated.

Soon after in 1998 the concept of screening a combinatorial library of polymeric materials was demonstrated [18]. A library of 112 degradable polyarylates was prepared by copolymerizing 14 different diphenols with 8 different aliphatic diacids, and the water contact angle (WCA), glass transition temperature, mechanical properties, and fibroblast attachment and proliferation were assessed for each material. This study demonstrated how screening a combinatorial library of materials could be used to develop structure–property relationships. Generally cell proliferation decreased with increasing WCA; however, cell proliferation was unaltered by changes in surface hydrophobicity for materials where methylene groups in the polymer backbone were substituted with an oxygen atom. This result suggested that fibroblast proliferation was sensitive to subtle chemical changes that are not associated with changes in WCA. The throughput of this study was limited by the sample format: polymers in the library were spin coated onto glass cover slips and assessed individually. To circumvent this constraint, the polymer microarray was developed, allowing hundreds to thousands of unique polymers to be presented on a single glass slide [19–21]. Typically materials are presented as 300–500 µm diameter spots, allowing for approximately 2000 materials to be presented simultaneously. Allowing for replicate measurements, over 600 unique polymer compositions can be screened in a single assay. A table listing the key advances in the use of material microarrays for biomaterials discovery are presented in Table 4.1.

4.2.1   Substrate Preparation

A number of factors must be considered when designing a polymer microarray and applying it to the high throughput discovery of biomaterials [22], which are summarized in Fig. 4.1. This includes the identity of the substrate, the coating on the substrate, the members of the polymer library, and the biological assay which can couple with the microarray.

4.2.1.1   Substrate Material

The key substrate requirement is the ease with which it can be applied to the biological assay and associated high throughput readout. The substrate of choice is a glass microscope slide due to the range of stage holders, scanners, and microscopes

Table 4.1 Description of key advances in biomaterial microarray technology, in chronological order

Advancement	Description of study	Reference
Demonstrate a microarray of materials	Investigation of the formation of a combinatorial library of ceramic materials in an array format	Xiang et al. (1995) [17]
Demonstrate screening a combinatorial library of polymers for biological performance as an approach to biomaterials design	Study of the proliferation of fibroblasts on a combinatorial library of polymers.	Brocchini et al., (1998) [47]
Produce a microarray of polymers for screening cell response. Demonstrate in situ polymerization	Study of stem cell attachment and differentiation on a polymeric library	Anderson et al. (2004) [19]
Combinatorial screen of biological polymers	5 ECM proteins were printed as 32 different combinations to assess the attachment of rat hepatocytes and mouse embryonic stem cells (mES)	Flaim et al. (2005) [38]
Demonstration of printing pre-synthesized polymers for formation of a materials microarray	Study the attachment of stem cells with polymer materials	Anderson et al. (2005) [20]
	Study attachment of human renal tubular cells	Tourniaire et al. (2006) [21]
Development of the high throughput surface characterization of polymer microarrays	Study chemical and wettability properties of a polymer microarray	Urquhart et al. (2007) [78]
	Assessment of the binding of 3 different proteins to 21 polymer spots using SPRi	Hook et al. (2009) [64]
	AFM screen of a 576 member polymer microarray, identifying materials with switchable topographies	Hook et al. (2011) [81]
Establish PLS as a method for modelling a univariate dataset with multivariate chemical information	Study comparing wettability with the chemical functionality of a polymer spot as measured by ToF-SIMS	Urquhart et al. (2008) [79]
Production of a polymer microarray using ink-jet printing	Formation of a hydrogel microarray	Zhang et al. (2008) [29]
Development of a high throughput 3D culture system	hMSCs were fixed into microwells within a PEGDA matrix. The combined influence of the cell adhesive peptides, RGDSP and IKVAV, was assessed	Jongpaiboonkit et al. (2008) [76, 77]
Production of a topographical array	2,176 unique topographical units derived from circles, isosceles triangles, and rectangles were produced and used to study the influence of topography on the bioactivity of hMSCs	Unadkat et al. (2011) [32]
Application of polymer microarrays to discover materials that resist bacterial attachment	Screen of 370 polyurethanes with two bacterial strains	Pernagallo et al. (2011) [45]
	Assessment of the attachment of three bacterial strains to >700 polymers using a multi-generation-screening methodology	Hook et al. (2012) [33]
Model biological properties of polymer library from calculated molecular descriptors	hEB adhesion to a 496-member polymer library was modelled using a nonlinear Bayesian neural network model	Epa et al. (2012) [85]

Base substrate
- Typically glass
- Needs to be compatible with analysis equiptment

Substrate coating
- Agarose, PEGMA and pHEMA have been used
- Adhesion layer for polymer spots
- Low-fouling

Polymer library
- In situ polymerisation or pre-synthesised
- Contact or ink-jet printing
- Design for maximum diversity or optimisation of composition
- Need to consider number of repeats and arrangement of those repeats

Biological assay
- Must have a high-throughput readout
- Usually utilises fluorescent marker
- Readout measured on fluorescence scanner or automated microscope

Fig. 4.1 The key aspects of polymer microarray design. Schematic depiction of a polymer microarray where each colored dot represents a unique polymer. The various aspects of the polymer microarray that should be considered when designing an array are highlighted

that have been adapted to this format. Furthermore, the slide is transparent, which allows cells growing on the slide to be easily assessed using light microscopy. The glass slide is also widely compatible with most cell culture methods, enabling a diverse range of biological assays to be applied. However, polymer microarrays can equally be applied to other materials, for example directly onto polystyrene-based tissue cultureware, which may enable the polymer microarray format to be accessible to a greater number of biological assays.

4.2.1.2 Substrate Coating

The surface chemistry of the underlying substrate plays an important role in the formation of a microarray as well as the success of subsequent bioassays.

The substrate coating must be both adherent to the materials printed onto it and resistant to the attachment of biomolecules and living cells (non-fouling) in order to optimize the signal-to-noise ratio of any biological assay and prevent cross-talk from one spot to another. A number of different surface coatings have been developed with these needs in mind, with emphasis generally placed on using cheap and robust-coating methodologies. One such example is the dip-coating of a commercially available epoxide functionalized slide into a poly(hydroxyethyl methacrylate) (pHEMA) solution [19, 20]. Dip-coating is a simple technique that can easily be achieved in any laboratory, making this modification approach accessible. pHEMA is an attractive coating as it is able to resist cell attachment as well as providing a matrix into which printed material can penetrate and physically entangle to improve the stability of the spots [19]. Agarose is an alternative coating that may be applied by dip-coating using commercial aminoalkylsilanated slides [21]. Agarose and pHEMA are intended to prevent cell attachment whilst being nontoxic. Poly(ethylene glycol) (PEG)-based coatings have also been widely used for producing low-fouling coatings and are widely used to effectively inhibit biomolecular adsorption [23]. In order to produce a PEG-modified surface, a methodology has been proposed by the groups of Griesser, Thissen, and Voelcker in which a PEG layer is reacted with an amine plasma polymer-coated slide [24–26]. Key to the success of a PEG coating is the production of a dense, brush-like layer. Polymers arrayed onto this surface can be covalently attached by modifying the polymer with a cross-linker or incorporating epoxy groups into the PEG layer. For example, a grafting-to approach was accomplished by producing a multifunctional coating with both PEG groups, which provided a low-fouling background, and epoxy groups, which enabled subsequently spotted biomolecules, synthetic and natural polymers to be covalently linked to the surface in a suitable manner for subsequent cell attachment assays using HeLa cells [26]. This is advantageous because the structure of the arrayed material is unaltered; however, the substrate chemistry becomes limited. In this particular study a plasma polymer slide-coating approach was used, which is advantageous as it can be applied to almost any base substrate.

4.2.2 Material Microarray Production

The creation of material arrays requires a combinatorial library of materials to be generated and positioned at addressable locations on a substrate surface. The library can be generated prior to the synthesis of the array, for example by printing pre-synthesized polymers, or by printing monomer solutions that are polymerized in situ. Once a material has been deposited onto the substrate it needs to adhere to the surface with sufficient strength and durability that it can resist the biological evaluation process. This may be achieved by physical entanglement, non-covalent or covalent interactions. A schematic depicting the printing of a polymer microarray is shown in Fig. 4.2.

Fig. 4.2 Schematic of the formation of a polymer microarray using contact printing with a quilled pin. Initially the pin is dipped into a well-containing monomer or polymer solution and then moved using a robot to the substrate to transfer the solution. The pin can be replaced with an ink-jet nozzle. On the right is a fluorescence image of a polymer microarray formed by this method. The fluorescence seen is due to autofluorescence of the resultant polymer spots

4.2.2.1 Microarray Formation

Polymer microarrays are typically formed by either contact or ink-jet printing. Contact printing involves the use of a robot moving a metallic pin, which is dipped into a solution and then spotted onto the substrate surface by making contact. The pin may either be solid or contain a groove analogous to an ink quill, such that the solution is drawn up into the pin and the spotted material is taken from this reservoir. The ink quill design has the advantage that more spots can be printed from a single dip in the solution. Contact printing is attractive for ease of transfer and the absence of small apertures in the system that can become blocked. The size and shape of the pin used is the determining factor in the resultant spot size. Contact printing was first used to produce an acrylate microarray on a pHEMA-coated slide by Anderson et al. using in situ polymerization [19]. This was achieved in five steps; mixing monomers at various ratios in a source plate, printing acrylate monomers with an initiator, activation of initiator upon UV irradiation, polymerization of the monomer, and finally removal of the solvent. This resulted in stable, covalently cross-linked polymer spots. By premixing the various monomers at a set ratio, a large polymer library of 576 materials was readily achieved. This approach decreases the time required for polymer library synthesis and microarray formation by combining these processes and also allows materials that cannot be printed, such as cross-linked polymers, to be included in the polymer library. However, the polymer synthesis conditions are likely to differ when materials are produced on a larger scale.

An alternative approach is to print a library of pre-synthesized materials [20, 21], which allows standard polymer characterization techniques such as gel permeation chromatography, nuclear magnetic resonance, and differential scanning calorimetry to be utilized without having to extract minute amounts of polymer from the slide. However, this method is restricted to polymers that can be solvated, and is considerably more laborious for the initial creation of the polymer library.

An alternative to contact-printing is ink-jet printing. In this approach a nozzle is used to draw up, then eject polymer solution onto the substrate at a defined droplet volume. The volume of the droplet and the surface energy of the substrate material and the printed solution determine the resulting spot size. This technique enables the precise control of the amount of material deposited and avoids contact with the surface. However, solutions of different viscosities and surface energies cannot easily be printed under similar conditions, limiting the number of different materials that can be included in a single printing run. In addition, the small orifices are susceptible to blockage by dust particles or aggregates of material [27]. This method can be cost-effective and readily accessible to most laboratories by using modified, commercially available ink-jet printers [28].

Ink-jet printing was first used to prepare a polymer array from individually deposited monomer for water soluble acrylamide monomers to form hydrogels [29]. Three monomers were deposited sequentially onto the same position, with a solution containing a catalyst to initiate the reaction being printed subsequently. This *drop-in-drop* mixing approach required that the solvent, water in this case, did not evaporate before the mixing was complete. The turbulence induced by the printing procedure resulted in the complete mixing of the monomers within the drops after 1.5 min. This approach was used to create an array comprising 36 different materials from six monomers. The polymers produced using the on-slide mixing methodology must be carefully assessed as they are likely to differ for each monomer, polymerization, and printing system employed.

4.2.2.2   The Design of the Polymer Library

When designing the components of a polymer microarray, careful consideration must be given to the objective of the experiment. In cases where the aim of an experiment is to *generate new data* for a poorly understood system the polymer library should be designed to *maximize the diversity* of chemical and/or physical properties represented. This can be achieved by selecting a large number of unique base components [30], but can equally be achieved by producing a combinatorial library of polymers by mixing a smaller number of base components at various ratios [31, 32]. In some cases sufficient understanding of the biological–material interaction can exist to allow the formulation of a hypothesis that can be tested by the judicious design of a polymer library [30, 33, 34]. This may include the *optimization* of a "hit" formulation identified from a previous screen, or testing a structure–function relationship. Typically these material libraries include subsets of polymers formed from a pair of base components where the composition of each component is varied systematically and sequentially. This results in material gradients that are formed from discrete units rather than a continuously changing material. It is prudent to randomize the position of compositionally similar materials (including replicates) on an array to avoid the occurrence of pseudo-trends that result from experimental conditions such as the inhomogeneous distribution of cells within a cell culture well. Gradients are highly useful tools for the optimization of materials, but have limited applicability to material discovery applications. Hansen et al. [35] combined gradients and microarrays, forming an array of polymer gradients. This enabled both the combinatorial screen of polymers and also compositional optimization. Successful formation of the arrays was assessed by X-ray photoelectron spectroscopy (XPS) and also through the introduction of a fluorescently tagged monomer, allowing the gradient to be tracked by fluorescence microscopy. Onto a single glass slide, 24 monomers were printed as 84 pairs to produce 84 polymer gradients. The array of gradients was used to assess the adhesion of HeLa and K562 cells.

Both design strategies, *maximizing diversity* or *optimizing a composition*, are complementary and can be utilized to employ a multi-generation approach to materials discovery whereby the first-generation array is designed to screen a large diversity of chemical and/or physical properties and subsequent generations seek to optimize hit formulations [33, 34].

4.3   Biological Application of Material Microarrays

The final consideration for applying polymer microarrays to the discovery of biomaterials is the design of a suitable bioassay that is compatible with the high throughput format. To achieve this, the key requirement is the inclusion of a suitable readout that can be assessed in an automated fashion. Typically, a fluorescence marker is used that can be measured using an automated fluorescence microscope or a fluorescence scanner. Utilizing this general strategy polymer microarrays have been successfully used to probe a diverse range of biological systems including supporting stem cell outgrowth [19, 31, 32, 36–43], maturation and phagocytosis of dendritic cells [44], materials resistant to bacteria [30, 33, 45], switchable materials [34, 43, 46], platelet activation [47], cell sorting [39, 48, 49], hepatocytes and toxicity models [38, 50], human skeletal cell attachment [49, 51], endothelization [52], *giardia lamblia* material interactions [53], cell transfection [13, 14, 54], and *Cryptosporidium parvum* material interactions [55].

The development of novel biomaterials has focussed on a number of key aspects, which are depicted schematically in Fig. 4.3. This includes probing cell–material interactions, including biological recognition of surface-bound biomolecules, the response of cells to surface chemistry and topography, as well as mechanical properties of the biomaterial and whether the cell culture system is 2D/3D (2/3 dimensions). The key studies that have explored these aspects are discussed below.

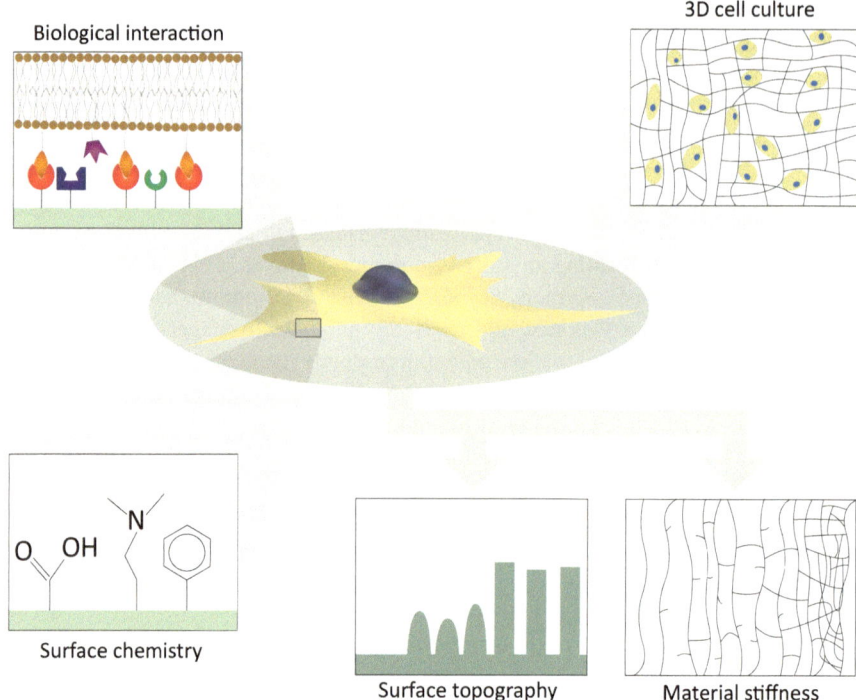

Fig. 4.3 Key aspects of biomaterials development explored using polymer microarrays. The use of polymer microarrays to discover novel biomaterials focusses on exploring cell–material interactions. The various aspects of these interactions that have been investigated are highlighted

4.3.1 Synthetic Polymers

In the pioneering study by Anderson et al. [19], a polymer microarray containing 496 unique polymers was produced from 24 monomers to screen for materials that support human embryonic stem cell (hES) attachment. This work demonstrated the large number of biological–material interactions that could be investigated in parallel. A number of different responses of hES cells derived from human embryoid bodies (hEB) were observed, including cell attachment, no attachment, and directed differentiation. Subsequent work targeted new materials that were able to support the clonal growth of hES in a xeno-free environment [31]. Stem cells hold enormous potential for regenerative medicine and tissue-engineering applications as they are able to differentiate into any cell within the body. However, methods to culture hES rely on animal products, typically using matrigel as a cell adhesive matrix or a layer of "feeder" mouse embryonic fibroblast cells. This limits the ability to expand these cells in vitro for clinical use. To identify xeno-free materials able to support stem cell attachment and expansion hES were cultured with an array of 496 materials produced by mixing 22 monomers at various ratios. Protein readily adsorbs to the surfaces of materials, thus, in most cell culture conditions the attachment of cells is regulated by the intermediate adsorbed protein layer [31]. Thus, the microarray was pre-adsorbed with laminin, fibronectin, fetal bovine serum, and bovine serum albumin to study how different proteins alter cell attachment. Fetal bovine serum (FBS)-coated materials resulted in the highest level of attachment without initiating cell differentiation, as determined by the highest number of cells expressing Oct4 (a marker of pluripotent cells) [56]. The "hit" materials identified typically contained a large number of acrylate groups suggesting the degree of cross-linking plays a role on cell attachment although a direct correlation with material hardness was not observed. For subsequent scaled-out experiments the FBS pre-adsorption was successfully replaced with a vitronectin pre-adsorption step and culture was conducted in mTeSR1, a fully defined medium, resulting in a xeno-free substrate for cell culture. A long-term cell culture was supported under these conditions for a month (five passages).

Zhang et al. [43] sought to develop a material that would allow stem cells to detach from a surface upon a reduction in temperature rather than using chemical, enzymatic or mechanical methods of removal. To achieve this, an array of 609 unique acrylate and acrylamide polymers was prepared composed of various mixtures of 18 monomer base units. Initially hits were selected that could support hES culture up to 7 days. The top 25 polymers were then scaled-up and assessed for their ability to release attached cells upon a reduction in temperature. The top three polymers that could support hES culture and achieved a thermoresponsive release all contained the monomers 2-(acryloyloxyethyl) trimethylammonium chloride and 2-(diethylamino)ethyl acrylate. On a copolymer of these two monomers hES were passaged 20 times utilizing a temperature reduction to release attached cells. Cell growth was typically slower on this polymer than compared to Matrigel. Cells grown on the thermoresponsive polymer were positive for Nanog and Oct3/4, which are markers for pluripotent stem cells.

Polymer microarrays have been applied to a number of other eukaryotic cells, for example dendritic cells [44]. These cells play a key role in regulating the immune response, and are targets for vaccine design. These cells are usually sourced as bone marrow dendritic cells (BMDCs), immature dendritic cells that are highly phagocytic, however, the cells are highly susceptible to maturation, preventing them from capturing antigens by phagocytosis, and are difficult to immobilize as this event can also cause their maturation. Effective immobilization of BMDCs whilst maintaining their immature state would enable the presentation of these cells for phenotypic studies, or allow cell-based assays that would lead to vaccine development. Mant et al. [44] used polymer microarrays to screen 120 polyurethanes to identify materials for the immobilization of BMDCs labelled with flurorescein, enabling a rapid fluorescent read-out of cell numbers. Hit polymers, all of which contained poly(tetramethylene glycol), were spin coated onto glass slides and attached cells were assessed for their ability to prevent maturation of the BMDCs by antibody staining for CD11c, a marker for this cell line. The initial BMDC culture was 90 % pure, and once attached the majority (approximately 70 %) of cells were maintained as BMDCs. Phagocytosis was also assessed of attached cells and compared with cells attached to poly-L-lysine, the traditional cell adhesive polymer coating. Cells were incubated with 3 μm latex microspheres and the degree of phagocytosis was assessed by counting the number of microspheres internalized by the cells. The maximum number of internalized microspheres was 3.6 per cell, compared with 1.3 for cells cultured on poly-L-lysine-coated samples. The authors noted improved cells attachment correlated with a decreased number of phagocytized microspheres, suggesting that the two processes share similar signalling pathways and are, thus, directly in competition.

In another example, Hansen et al. [47] used polymer microarrays to search for a synthetic polymer that would activate platelets to cause hemostasis to replace costly naturally derived products including materials based upon fibrin, chitosan, or cellulose. This is towards the goal of improving wound treatment by providing materials that cause the rapid and effective hemostasis of injuries. In this study 291 unique polyacrylates were incubated with platelet rich plasma for 30 min, after which the attached platelets were stained with antibodies CD41-FITC, which recognizes GPIIb a protein present on all platelets [57], and CD62P-PE, which is a marker specific to activated platelets [58]. In this experiment the binding of platelets to a material was synonymous with platelet activation. The "hit" polymer showed a fourfold higher activation compared with collagen. All "hit" polymers contained a sterically non-hindered tertiary amine, suggesting that surface charge may play a role in successfully activating platelets.

In addition to studies investigating the attachment and interaction of eukaryotic cells with polymers, material microarrays have also been used to discover novel polymers that resist the attachment of prokaryotic cells [30, 33, 45]. Bacterial attachment to biodevices can lead to the formation of biofilms, within which bacteria are 1,000 times more resistant to the host immune system and antibiotics [59]. Biofilms are estimated to be associated with 80 % of hospital acquired infections [60]. It is, therefore, of interest to discover new materials that can be used as biodevices to prevent bacterial attachment, subsequent biofilm formation and hence reduce the incidence of device-associated infections. Hook et al. [33] screened a library of 496 unique polymers to identify materials that could resist the attachment of *Pseudomonas aeruginosa*, *Staphylococcus aureus*, and *Escherichia coli*. Each of these strains was transfected with a plasmid encoding for a green fluorescing protein in order to provide a rapid fluorescent readout of bacterial numbers using a fluorescence scanner. This study used a multiple generation approach, whereby bacterial attachment was initially assessed on the first-generation array designed to screen a large chemical variance, whilst the second-generation array sought to optimize the composition of "hit" polymers. Using this approach, a new class of materials resistant to bacteria attachment was discovered. This underlines the importance of high throughput screening studies, where new material–biological interactions can be uncovered that would not be predicted from the current understanding. The new class of materials consisted of a hydrophilic ester group combined with cyclic hydrocarbon pendant groups to create weak amphiphiles. This is significantly different from the alternative anti-adhesive materials (PEG and zwitterions) that are highly hydrophilic and are able to resist bacteria on the basis of exclusion caused by a highly organized water layer [61–63]. The weakly amphiphilic polymer was able to reduce coverage of bacteria by up to 96 % compared with silicone, and significantly, when scaled up and applied as a coating the anti-adhesive polymer was able to reduce bacteria numbers by an order of magnitude in vivo in a mouse model. An exhaustive search of the polyacrylate combinatorial space was conducted in a subsequent study that again utilized a multiple generation approach [30]. One hundred and sixteen monomers were used to produce 1,273 unique materials and in over 10,000 separate assays an optimal antibacterial polymer was identified with up to 99 % reduction in bacterial coverage compared with silicone. This study highlighted the efficacy of the new class of weakly amphiphilic polymers at resisting bacterial attachment.

4.3.2 Biological Polymers

High throughput screens of material–cell interactions are not limited to synthetic polymers but can equally be applied to biological polymers, and can offer greater control of cells due to the inherent biological activity of the polymers. These studies have focussed on cell adhesive molecules such as extracellular matrix (ECM) proteins and cell adhesive-derived biomolecules. Pioneering studies were performed by Flaim et al. [38]. In this study, collagen I, collagen III, collagen IV, laminin, and fibronectin were printed as 32 different combinations onto a hydrogel-coated glass slide to assess the attachment of rat hepatocytes and mouse embryonic stem cells (mES). After hepatocyte attachment cells were immunostained for albumin as a marker of liver-specific function. The highest albumin signals were associated with ECM combinations that contained collagen IV. However, albumin signal did not correlate with the amount of collagen when hepatocytes were cultured on serially diluted collagen IV without other ECM components. This result showed that two ECM components could act synergistically to produce an effect that would not be predicted if the components were assessed separately [38]. As another example, Fn and collagen III individually induced a negative effect on hepatocyte function compared to the average response to other ECM components present on the array, however, when combined with collagen I a positive effect on hepatocyte function was observed. The occurrence of synergistic effects is a key justification for the use of combinatorial libraries of polymers for high throughput screens. When applied to mES, the combination of collagen I and Fn was best able to instigate the differentiation of ES cells to an early hepatic fate, as observed by increased levels of B-galactosidase. This demonstrated the possibility of screening ES cells with a materials microarray to identify surfaces that trigger differentiation pathways in a controlled manner. The stem cell niche was probed further by Gobaa et al. [40] by printing biomolecules directly onto silicon pillars that were subsequently used as a microstamp to transfer the biomolecule pattern onto a PEG hydrogel. Simultaneously the hydrogel was embossed to create microwells that were 35 μm deep and 450 μm in diameter. Biomolecules were tethered to the bottom of the wells using either a nonspecific cross-linker or through the interaction of an Fc tag with protein A or G.

The stiffness of the gel was easily altered by changing the amount of cross-linker, allowing for both biomolecule composition and material stiffness to be assessed in parallel. This platform was used to screen an array of 23 candidate proteins with mouse neural stem cells. This study identified Jagged 1 and DLL4 as the main inducers of neurosphere growth, consistent with the previous literature.

A further aim of high throughput material screens has been to identify specific cell–material interactions that can allow certain cell types to be selected from mixed populations. One specific area of investigation has been the isolation of spermato-gonia stem cells from primary sources [26, 39, 48, 64]. These cells are pluripotent [65] and can be used for germ cell transfer technologies, however, little is known of their cell surface markers that could be used for their isolation from primary sources and subsequent culture. A microarray was formed from 27 biological factors that represented a range of potential germ cell-recognition properties. This included ECM proteins, cell-surface active growth factors, lectins, antibodies against cell surface proteins, as well as amine-functional synthetic polymers. The microarrays were incubated with mixed cell populations derived from freshly isolated bovine testicular tissue. Cells positive for Plzf (a nuclear transcription factor present in type A spermatogonia [66]) were detected by immunohistochemistry. A vimentin anti-body was used to identify Sertoli and myoid cell populations [67]. The lectin *Dolichos biflorus* was identified to successfully enrich spermatogonia stem cells from sertoli and myoid cells, whereas the lectins *Pisum sativum* and Concanavalin A preferentially attached sertoli and myoid cells over the spermatogonia stem cells, thus, may be useful to enrich spermatogonia stem cells through negative selection.

4.3.3 Topography

The topography of a surface plays a key role in the attachment and behavior of cells; contact guidance of cells along grooves has been observed down to nanoscale fea-tures [68, 69], altering the roughness of prosthetic implants has been a key param-eter to improve bone-implant contact [70–72], and it has also been demonstrated that surface roughness can influence the differentiation of preosteoblast cells [73]. Thus, surface topography is a key parameter to be included in biomaterial develop-ment programs and, as such, high throughput tools for screening surface topography will be invaluable for furthering the understanding of cell–material interactions.

Unadkat et al. [32] reported on a high throughput platform for screening the response of cells to various topographic features. Initially, 2,176 unique topographi-cal units derived from circles, isosceles triangles, and rectangles were designed using mathematical algorithms that arranged these three base shapes within an imaginary square. The three base shapes were selected because they allow the inclu-sion of large smooth areas (circles), angles (triangles), and stretched elements (rect-angles) within the resultant topographical features. A silicon mould of these patterns was produced by photolithography and used to emboss poly(lactic acid) films. These films were used to study the influence of topography on the bioactivity of human mesenchymal stem cells (hMSCs). Varying levels of alkaline phosphatase, which is a marker for early osteogenic differentiation [74], were observed for cells grown on the varied topographical patterns demonstrating that various topographies can induce cellular differentiation. Increased cell spreading was accompanied with osteogenic differentiation.

In another strategy to investigate material topography, Adler et al. [54] produced an array of pits, either square or circular, using chemical etching through a resist. Patterns were made with poly(dimethyl siloxane) patterned from a metallic master. Ten patterns of pits were produced with 16 variations in pit spacing and size, resulting in 160 variations in surface topography. This materials array was used to assess the interaction of topography on the transfection of fibroblast cells. The improved efficiency of nonviral transfection is an important development to advance gene medicines. Cells were attached to the patterned substrate and transfection efficiencies were assessed after 24 h. Up to 25 % enhancement in transfection efficiency was observed for cells grown on pitted topographies compared with a flat surface. Large pits close together offered the greatest improvement, but also produced the least spread cells.

4.3.4 3D Cell Culture

A key challenge for tissue-engineering applications is translating the 2D in vitro culture methods into 3D devices in vivo. Towards this goal 3D in vitro culture methods have been developed [75] that attempt to bridge the gap between in vitro and in vivo cell culture. Concurrently, 3D arrays have been developed that enable multiple experimental factors to be explored within 3D matrices in parallel.

Jongpaiboonkit et al. [76] developed an array of 3D culture matrices based upon a PEG diacrylate background that contained microwells. A liquid-handling system filled these wells with a hydrogel precursor solution that also contained cells of varied density. The hydrogels were then UV cured to fix the cells within the 3D matrix. The entire array could then be placed into a larger multi-well plate for culture. This system was used to study the interaction of the cell adhesive peptide RGDSP and growth factor (FGF2) on hMSCs [42]. The viability of cells grown in the wells was maintained at 82.7 % after 7 days growth with RGDSP and FGF2 compared with 45.5 % without these added components. The array format allowed for the optimal combination of these biological factors to be determined. The combined influence of the cell adhesive peptides, RGDSP and IKVAV on hMSCs, was also assessed [77]. After 7 days culture within the 3D matrices, the viability was improved from 45 % to up to 70 % with the addition of cell adhesive peptides. However, viability was improved further up to 80 % when both peptides were used.

In a different approach, Fernandes et al. [41] printed mouse EBs in an alginate gel to produce a 3D cell culture array. Cells remained viable, undifferentiated, and expanded within the 3D matrix. As a proof-of-concept of the high throughput utility of this system, a small molecule array containing FGF-4 or reionoic acid was stamped onto the 3D cell culture and the small molecules diffused into the cultured cells. These molecules induced cellular differentiation as evidenced by a drop in Oct-4 and Nanog expression, markers of pluripotency.

4.3.5   Development of Structure–Function Relationships

A key advantage of high throughput screening methodologies is the discovery of new materials optimal for a given biological application. Additionally, the large number of biological–material interactions assessed using polymer microarrays can be used to develop structure–function relationships that, in turn, allow the design of biologically functional materials. However, this requires extensive surface characterization of the materials within a library. High throughput measurements of polymer microarrays have been achieved by XPS, time-of-flight secondary ion mass spectrometry (ToF-SIMS) [78, 79], WCA [80], atomic force microscopy (AFM) [81], surface plasmon resonance (SPR) [64], and force measurements [82]. The application of AFM and force measurements to a polymer microarray allows the mechanical properties of the materials to be included in subsequent structure–function relationships [31]. "High throughput surface characterization" refers to the automation of measurements, allowing an instrument to be set up to take measurements from every member of a library with little to no user input. Once material properties have been measured the challenge remains to correlate these properties with a biological response. This is particularly difficult because the systems typically being studied are complex and poorly understood. Simple correlations with properties that are commonly implicated in cell–material interactions, such as WCA or roughness, typically do not work. The key step in achieving a correlation between biological response and a measured surface property has been the use of partial least square (PLS) regression to correlate a univariate property, such as cell numbers, with a multivariate dataset, such as the hundreds of secondary ions produced in ToF-SIMS spectra [83]. This method was initially validated by linking ToF-SIMS spectra with WCA [83], and has been successfully applied to predict the frequency of colony formation of stem cells from the chemical information represented in ToF-SIMS spectra [31, 37]. Similarly bacterial attachment was predicted for *P. aeruginosa* and *S. aureus* [33]. The successful construction of these models likely results from the diverse material properties that are represented within ToF-SIMS spectra. Secondary ions emitted from a surface can be assigned to chemical functionalities that are associated with surface charge, hydrophobicity, aromatic and cyclic groups, and material stiffness [31, 33, 34, 79, 83]. However, the PLS models produced have thus far been restricted to interpolative predictions. A powerful application of this approach would enable the design of hit materials based upon the models produced. Limiting this is the extensive surface analysis required to gather the relevant chemical/physical properties of the materials being studied. Moreover, the number of materials that can be included is limited to the number of different samples that can feasibly be screened. Although this is significantly expanded by the use of polymer microarrays to the order of hundreds to thousands of materials, this is still small compared with the infinite number of varied polymer chemistries that are possible.

To overcome this constraint, modelling based upon chemical descriptors rather than a measured property has been explored [84]. Using this method, materials do not need to be produced and analyzed, but rather a virtual library of materials can be created where the size of the library is limited by computational power rather than experimental requirements. Hook et al. used this approach to predict defects within polymers from their molecular descriptors using a PLS regression model [84]. Materials were initially screened by ToF-SIMS imaging and light microscopy to identify defects such as chemical heterogeneity or spreading. A PLS model based upon the molecular descriptor was successfully able to predict whether a pair of monomers would produce a defective polymer in 85 % of cases. This study demonstrated that the molecular descriptors of a polymer and their material properties are linked, and as such other material properties may be predicted.

This work was expanded by Epa et al. [85], whereby nonlinear Bayesian neural network models were developed that could predict the adhesion of hEB using only calculated molecular descriptors. In particular, this work demonstrated substantially higher predictive power of nonlinear models compared with linear models. An r^2 value from a plot of measured versus predicted values was reported as 0.68 for a linear model with a standard error estimation (SEE) of 0.163log EB whereas for a nonlinear model the r^2 value was 0.81 with an SEE of 0.108log EB. This suggests that there is some nonlinearity in the relationship between hEB adhesion and polymer structure [85]. Combining such modelling methods with polymer microarray screens, and specifically the design of polymer libraries, enables exciting opportunities to assess a larger material property space than could be accessed by experimental methods alone.

4.4   Closing Remarks

Polymer microarrays are a key enabling technology for biomaterials discovery. They have been successfully used to develop the surface chemistry, topography, bioactivity, and mechanical properties of biomaterials as well as allowing the high throughput study of 3D matrices. Polymer microarrays have been applied to numerous biological applications including supporting cell attachment and outgrowth, maturation and phagocytosis of dendritic cells, materials resistant to microbes, switchable materials, platelet activation, cell sorting, hepatocytes and toxicity models, and cell transfection. Additional to the discovery and development of new biomaterials, the large datasets generated by polymer microarrays can also be used to develop structure–function relationships and help elucidate underlying biological–material interactions. In particular, the modelling of the biological response of polymers using calculated molecular descriptors enables the screening of virtual libraries of polymers. This will allow a larger set of material properties to be accessed than by experimental methods alone. Future insight into material–cell interactions and expanded knowledge of currently *ideas-poor* fields of material science will be greatly aided by polymer microarrays and their application to increasingly ambitious biological assays and methods for presenting materials.

References

1. Kell DB, Oliver SG (2003) Here is the evidence, now what is the hypothesis? The complementary roles of inductive and hypothesis-driven science in the post-genomic era. Bioessays 26:99–105. doi:10.1002/bies.10385
2. Derda R, Musah S, Orner BP, Klim JR, Li LY, Kiessling LL (2010) High-throughput discovery of synthetic surfaces that support proliferation of pluripotent cells. J Am Chem Soc 132:1289–1295. doi:10.1021/ja906089g
3. Bleicher KH, Bohm HJ, Muller K, Alanine AI (2003) Hit and lead generation: beyond high-throughput screening. Nat Rev Drug Discov 2:369–378. doi:10.1038/nrd1086
4. Gallop MA, Barrett RW, Dower WJ, Fodor SPA, Gordon EM (1994) Application of combinatorial technologies to drug discovery. 1. Background and peptide combinatorial libraries. J Med Chem 37:1233–1251. doi:10.1021/jm00035a001
5. Gordon EM, Barrett RW, Dower WJ, Fodor SPA, Gallop MA (1994) Applications of combinatorial technologies to drug discovery. 2. Combinatorial organic-synthesis, library screening strategies, and future-directions. J Med Chem 37:1385–1401. doi:10.1021/jm00036a001
6. Heller MJ (2002) DNA microarray technology: devices, systems, and applications. Annu Rev Biomed Eng 4:129–153. doi:10.1146/annurev.bioeng.4.020702.153438
7. Kumar A, Goel G, Fehrenbach E, Puniya AK, Singh K (2005) Microarrays: the technology, analysis and application. Eng Life Sci 5:215–222. doi:10.1002/elsc.200420075
8. Leung YF, Cavalieri D (2003) Fundamentals of cDNA microarray data analysis. Trends Genet 19:649–659. doi:10.1016/j.tig.2003.09.015
9. Stoll D, Templin MF, Bachmann J, Joos TO (2005) Protein microarrays: applications and future challenges. Curr Opin Drug Dis Dev 8:239–252
10. Templin MF, Stoll D, Schrenk M, Traub PC, Vohringer CF, Joos TO (2002) Protein microarray technology. Trends Biotechnol 20:160–166. doi:10.1016/S0167-7799(01)01910-2
11. Wilson DS, Nock S (2003) Recent developments in protein microarray technology. Angew Chem Int Edit 42:494–500. doi:10.1002/anie.200390150
12. Angres B (2005) Cell microarrays. Expert Rev Mol Diagn 5:769–779. doi:10.1586/14737159.5-5.769
13. Hook AL, Creasey R, Thissen H, Hayes JP, Voelcker NH (2009) Laser-based patterning for transfected cell microarrays. Biofabrication 1:045003. doi:10.1088/1758-5082/1/4/045003
14. Hook AL, Thissen H, Voelcker NH (2009) Advanced substrate fabrication for cell microarrays. Biomacromolecules 10:573–579. doi:10.1021/bm801217n
15. Hook AL, Voelcker N, Thissen H (2006) Surface manipulation of biomolecules for cell microarray applications. Trends Biotech 24:471–477. doi:10.1016/j.tibtech.2006.08.001
16. Ziauddin J, Sabatini DM (2001) Microarrays of cells expressing defined cDNAs. Nature 411:107–110. doi:10.1038/35075114
17. Xiang XD, Sun XD, Briceno G, Lou YL, Wang KA, Chang HY, Wallacefreedman WG, Chen SW, Schultz PG (1995) A combinatorial approach to materials discovery. Science 268:1738–1740. doi:10.1126/science.268.5218.1738
18. Brocchini S, James K, Tangpasuthadol V, Kohn J (1998) Structure–property correlations in a combinatorial library of degradable biomaterials. J Biomed Mater Res 42:66–75. doi:10.1002/(SICI)1097-4636(199810)42:1<66::AID-JBM9>3.0.CO;2-M

19. Anderson DG, Levenberg S, Langer R (2004) Nanoliter-scale synthesis of arrayed biomaterials and application to human embryonic stem cells. Nat Biotech 22:863–866. doi:10.1038/nbt981
20. Anderson DG, Putnam D, Lavik EB, Mahmood TA, Langer R (2005) Biomaterial microarrays: rapid, microscale screening of polymer-cell interaction. Biomaterials 26:4892–4897. doi:10.1016/j.biomaterials.2004.11.052
21. Tourniaire G, Collins J, Campbell S, Mizomoto H, Ogawa S, Thaburet JF, Bradley M (2006) Polymer microarrays for cellular adhesion. Chem Comm 28:2118–2120. doi:10.1039/b602009g
22. Kohn J, Welsh WJ, Knight D (2007) A new approach to the rationale discovery of polymeric biomaterials. Biomaterials 28:4171–4177. doi:10.1016/j.biomaterials.2007.06.022
23. Kingshott P, Thissen H, Griesser HJ (2002) Effects of cloud-point grafting, chain length, and density of PEG layers on competitive adsorption of ocular proteins. Biomaterials 23:2043–2056
24. Hook AL, Thissen H, Hayes JP, Voelcker NH (2005) Spatially controlled electro-stimulated DNA adsorption and desorption for biochip applications. Biosens Bioelectron 21:2137–2145. doi:10.1016/j.bios.2005.10.008
25. Yasuda H (1985) Plasma polymerization. Academic, Orlando
26. Kurkuri MD, Driever C, Johnson G, McFarland G, Thissen H, Voelcker NH (2009) Multifunctional polymer coatings for cell microarray applications. Biomacromolecules 10:1163–1172. doi:10.1021/bm801417s
27. Tekin E, de Gans BJ, Schubert US (2004) Ink-jet printing of polymers—from single dots to thin film libraries. J Mater Chem 14:2627–2632. doi:10.1039/b407478e
28. Pardo L, Wilson WC, Boland TJ (2003) Characterization of patterned self-assembled monolayers and protein arrays generated by the ink-jet method. Langmuir 19:1462–1466. doi:10.1021/la026171u
29. Zhang R, Liberski A, Khan F, Diaz-Mochon JJ, Bradley M (2008) Inkjet fabrication of hydrogel microarrays using in situ nanolitre-scale polymerisation. Chem Comm 11:1317–1319. doi:10.1039/b717932d
30. Hook AL, Chang C, Yang J, Atkinson S, Langer R, Anderson DG, Davies MC, Williams P, Alexander MR (2013) Discovery of novel materials with broad resistance to bacterial attachment using combinatorial polymer microarrays. Adv Mater 25:2542–2547. doi:10.1002/adma.201204936
31. Mei Y, Saha K, Bogatyrev SR, Yang J, Hook AL, Kalcioglu ZI, Cho SW, Mitalipova M, Pyzocha N, Rojas F, Van Vliet KJ, Davies MC, Alexander MR, Langer R, Jaenisch R, Anderson DG (2010) Combinatorial development of biomaterials for clonal growth of human pluripotent stem cells. Nat Mater 9:768–778. doi:10.1038/nmat2812
32. Unadkat HV, Hulsman M, Cornelissen K, Papenburg BJ, Truckenmuller RK, Carpenter AE, Wessling M, Post GF, Uetz M, Reinders MJT, Stamatialis D, van Blitterswijk CA, de Boer J (2011) An algorithm-based topographical biomaterials library to instruct cell fate. Proc Natl Acad Sci U S A 108:16565–16570. doi:10.1073/pnas.1109861108
33. Hook AL, Chang C, Yang J, Luckett J, Cockrayne A, Atkinson S, Mei Y, Bayston R, Irvine DJ, Langer R, Anderson DG, Williams P, Davies MC, Alexander MR (2012) Combinatorial discovery of polymers resistant to bacterial attachment Nat Biotech 30:868–875. doi:10.1038/nbt.2316
34. Hook AL, Scurr DJ, Anderson DG, Langer R, Williams P, Davies M, Alexander MR (2013) High throughput discovery of thermo-responsive materials using water contact angle measurements and time-of-flight secondary ion mass spectrometry. Surf Interface Anal 45:181–184. doi:10.1002/sia.4910
35. Hansen A, Zhang R, Bradley M (2012) fabrication of arrays of polymer gradients using inkjet printing. Macromol Rapid Comm 33:1114–1118. doi:10.1002/marc.201200193
36. Mei Y, Gerecht S, Taylor M, Urquhart A, Bogatyrev SR, Cho SW, Davies MC, Alexander MR, Langer RS, Anderson DG (2009) Mapping the Interactions among biomaterials, adsorbed proteins, and human embryonic stem cells. Adv Mater 21:2781–2786. doi:10.1002/adma.200803184
37. Yang J, Mei Y, Hook AL, Taylor M, Urquhart AJ, Bogatyrev SR, Langer R, Anderson DG, Davies MC, Alexander MR (2010) Polymer surface functionalities that control human embryoid body cell adhesion revealed by high throughput surface characterization of combinatorial material microarrays. Biomaterials 31:8827–8838. doi:10.1016/j.biomaterials.2010.08.028

38. Flaim CJ, Chien S, Bhatia SN (2005) An extracellular matrix microarray for probing cellular differentiation. Nat Method 2:119–125. doi:10.1038/NMETH736
39. Anglin E, Davey R, Herrid M, Hope S, Kurkuri M, Pasic P, Hor M, Fenech M, Thissen H, Voelcker NH (2010) Cell microarrays for the screening of factors that allow the enrichment of bovine testicular cells. Cytom Part A 77A:881–889. doi:10.1002/cyto.a.20913
40. Gobaa S, Hoehnel S, Roccio M, Negro A, Kobel S, Lutolf MP (2011) Artificial niche microarrays for probing single stem cell fate in high throughput. Nat Method 8:949–955. doi:10.1038/nmeth.1732
41. Fernandes TG, Kwon SJ, Bale SS, Lee MY, Diogo MM, Clark DS, Cabral JMS, Dordick JS (2010) Three-dimensional cell culture microarray for high-throughput studies of stem cell fate. Biotech Bioeng 106:106–118. doi:10.1002/bit.22661
42. King WJ, Jongpaiboonkit L, Murphy WL (2010) Influence of FGF2 and PEG hydrogel matrix properties on hMSC viability and spreading. J Biomed Mater Res Part A 93A:1110–1123. doi:10.1002/jbm.a.32601
43. Zhang R, Mjoseng HK, Hoeve MA, Bauer NG, Pells S, Besseling R, Velugotla S, Tourniaire G, Kishen REB, Tsenkina Y, Armit C, Duffy CRE, Helfen M, Edenhofer F, de Sousa PA, Bradley M (2013) A thermoresponsive and chemically defined hydrogel for long-term culture of human embryonic, stem cells. Nat Comun 4:1335. doi:10.1038/ncomms2341
44. Mant A, Tourniaire G, Diaz-Mochon JJ, Elliott TJ, Williams AP, Bradley M (2006) Polymer microarrays: identification of substrates for phagocytosis assays. Biomaterials 27:5299–5306. doi:10.1016/j.biomaterials.2006.04.040
45. Pernagallo S, Wu M, Gallagher MP, Bradley M (2011) Colonising new frontiers-microarrays reveal biofilm modulating polymers. J Mater Chem 21:96–101. doi:10.1039/c0jm01987a
46. Zhang R, Liberski A, Sanchez-Martin R, Bradley M (2009) Microarrays of over 2000 hydrogels—identification of substrates for cellular trapping and thermally triggered release. Biomaterials 30:6193–6201. doi:10.1016/j.biomaterials.2009.07.055
47. Hansen A, McMillan L, Morrison A, Petrik J, Bradley M (2011) Polymers for the rapid and effective activation and aggregation of platelets. Biomaterials 32:7034–7041. doi:10.1016/j.biomaterials.2011.06.001
48. Anglin EJ, Salisbury C, Bailey S, Hor M, Macardle P, Fenech M, Thissen H, Voelcker NH (2010) Sorted cell microarrays as platforms for high-content informational bioassays. Lab Chip 10:3413–3421. doi:10.1039/c0lc00185f
49. Tare RS, Khan F, Tourniaire G, Morgan SM, Bradley M, Oreffo ROC (2009) A microarray approach to the identification of polyurethanes for the isolation of human skeletal progenitor cells and augmentation of skeletal cell growth. Biomaterials 30:1045–1055. doi:10.1016/j.biomaterials.2008.10.038
50. Hay DC, Pernagallo S, Diaz-Mochon JJ, Medine CN, Greenhough S, Hannoun Z, Schrader J, Black JR, Fletcher J, Dalgetty D, Thompson AI, Newsome PN, Forbes SJ, Ross JA, Bradley M, Iredale JP (2011) Unbiased screening of polymer libraries to define novel substrates for functional hepatocytes with inducible drug metabolism. Stem Cell Res 6:92–102. doi:10.1016/j.scr.2010.12.002
51. Khan F, Tare RS, Kanczler JM, Oreffo ROC, Bradley M (2010) Strategies for cell manipulation and skeletal tissue engineering using high-throughput polymer blend formulation and microarray techniques. Biomaterials 31:2216–2228. doi:10.1016/j.biomaterials.2009.11.101
52. Pernagallo S, Tura O, Wu M, Samuel K, Diaz-Mochon JJ, Hansen A, Zhang R, Jackson M, Padfield GJ, Hadoke PWF, Mills NL, Turner ML, Iredale JP, Hay DC, Bradley M (2012) Novel biopolymers to enhance endothelialisation of intra-vascular devices. Adv Healthc Mater 1:646–656. doi:10.1002/adhm.201200130
53. Pickering H, Wu M, Bradley M, Bridle H (2012) Analysis of Giardia lamblia interactions with polymer surfaces using a microarray approach. Environ Sci Technol 46:2179–2186. doi:10.1021/es203637e
54. Adler AF, Speidel AT, Christoforou N, Kolind K, Foss M, Leong KW (2011) High-throughput screening of microscale pitted substrate topographies for enhanced nonviral transfection efficiency in primary human fibroblasts. Biomaterials 32:3611–3619. doi:10.1016/j.biomaterials.2011.01.040

55. Wu M, Bridle H, Bradley M (2012) Targeting Cryptosporidium parvum capture. Water Res 46:1715–1722. doi:10.1016/j.watres.2011.12.041

56. Nichols J, Zevnik B, Anastassiadis K, Niwa H, Klewe-Nebenius D, Chambers I, Scholer H, Smith A (1998) Formation of pluripotent stem cells in the mammalian embryo depends on the POU transcription factor. Cell 95:379–391. doi:10.1016/s0092-8674(00)81769-9

57. McMichael AJ (1987) Leucocyte typing III: white cell differentiation antigens. Oxford University Press, Oxford

58. Hsulin SC, Berman CL, Furie BC, August D, Furie B (1984) A platlet membrane-protein expressed during platlet activation and secretion—studies using monoclonal-antibody specific for thrombin-activated platelets. J Bio Chem 259:9121–9126

59. Costerton JW, Stewart PS, Greenberg EP (1999) Bacterial biofilms: a common cause of persistent infections. Science 284:1318–1322. doi:10.1126/science.284.5418.1318

60. Davies D (2003) Understanding biofilm resistance to antibacterial agents. Nat Rev Drug Discov 2:114–122. doi:10.1038/nrd1008

61. Kingshott P, Wei J, Bagge-Ravn D, Gadegaard N, Gram L (2003) Covalent attachment of poly(ethylene glycol) to surfaces, critical for reducing bacterial adhesion. Langmuir 19: 6912–6921. doi:10.1021/la034032m

62. Cheng G, Li GZ, Xue H, Chen SF, Bryers JD, Jiang SY (2009) Zwitterionic carboxybetaine polymer surfaces and their resistance to long-term biofilm formation. Biomaterials 30: 5234–5240. doi:10.1016/j.biomaterials.2009.05.058

63. Jiang SY, Cao ZQ (2010) Ultralow-fouling, functionalizable, and hydrolyzable zwitterionic materials and their derivatives for biological applications. Adv Mater 22:920–932. doi:10.1002/adma.200901407

64. Hook AL, Thissen H, Voelcker NH (2009) Surface plasmon resonance imaging of polymer microarrays to study protein-polymer interactions in high throughput. Langmuir 25:9173–9181. doi:10.1021/la900735n

65. Guan K, Nayernia K, Maier LS, Wagner S, Dressel R, Lee JH, Nolte J, Wolf F, Li MY, Engel W, Hasenfuss G (2006) Pluripotency of spermatogonial stem cells from adult mouse testis. Nature 440:1199–1203. doi:10.1038/nature04697

66. Costoya JA, Hobbs RM, Barna M, Cattoretti G, Manova K, Sukhwani M, Orwig KE, Wolgemuth DJ, Pandolfi PP (2004) Essential role of Plzf in maintenance of spermatogonial stem cells. Nat Genet 36:653–659. doi:10.1038/ng1367

67. Steger K, Schimmel M, Wrobel KH (1994) Immunocytochemical demonstration of cytoskeletal proteins in seminiferous tubules of adult rams and bulls. Arch Histol Cytol 57:17–28. doi:10.1679/aohc.57.17

68. Johansson F, Carlberg P, Danielsen N, Montelius L, Kanje M (2006) Axonal outgrowth on nano-imprinted patterns. Biomaterials 27:1251–1258. doi:10.1016/j.niomaterials. 2005.07.047

69. Tessier-Lavigne M, Goodman CS (1996) The molecular biology of axon guidance. Science 274:1123–1133. doi:10.1126/science.274.5290.1123

70. Cochran DL, Buser D, ten Bruggenkate CM, Weingart D, Taylor TM, Bernard JP, Peters F, Simpson JP (2002) The use of reduced healing times on ITI (R) implants with a sandblasted and acid-etched (SLA) surface: early results from clinical trials on ITI (R) SLA implants. Clin Oral Implant Res 13:144–153. doi:10.1034/j.1600-0501.2002.130204.x

71. Martin JY, Schwartz Z, Hummert TW, Schraub DM, Simpson J, Lankford J, Dean DD, Cochran DL, Boyan BD (1995) Effect of titanium surface-roughness on proliferation, differentiation, and protein-synthesis of human osteoblast-like cells (MG63). J Biomed Mater Res 29:389–401. doi:10.1002/jbm.820290314

72. Price RL, Ellison K, Haberstroh KM, Webster TJ (2004) Nanometer surface roughness increases select osteoblast adhesion on carbon nanofiber compacts. J Biomed Mat Res Part A 70A:129–138. doi:10.1002/jbm.a.30073

73. Schneider GB, Zaharias R, Seabold D, Keller J, Stanford C (2004) Differentiation of preosteoblasts is affected by implant surface microtopographies. J Biomed Mater Res Part A 69A:462–468. doi:10.1002/jbm.a.30016

74. Jaiswal N, Haynesworth SE, Caplan AI, Bruder SP (1997) Osteogenic differentiation of purified, culture-expanded human mesenchymal stem cells in vitro. J Cell Biochem 64:295–312. doi:10.1002/(sici)1097-4644(199702)64:2<295::aid-jcb12>3.0.co;2-i

75. Cukierman E, Pankov R, Stevens DR, Yamada KM (2001) Taking cell-matrix adhesions to the third dimension. Science 294:1708–1712. doi:10.1126/science.1064829

76. Jongpaiboonkit L, King WJ, Lyons GE, Paguirigan AL, Warrick JW, Beebe DJ, Murphy WL (2008) An adaptable hydrogel array format for 3-dimensional cell culture and analysis. Biomaterials 29:3346–3356. doi:10.1016/j.biomaterials.2008.04.040

77. Jongpaiboonkit L, King WJ, Murphy WL (2009) Screening for 3D environments that support human mesenchymal stem cell viability using hydrogel arrays. Tissue Eng Part A 15:343–353. doi:10.1089/ten.tea.2008.0096

78. Urquhart AJ, Anderson DG, Taylor M, Alexander MR, Langer R, Davies MC (2007) High throughput surface characterisation of a combinatorial material library. Adv Mater 19:2486–2491. doi:10.1002/adma.200700949

79. Urquhart AJ, Taylor M, Anderson DG, Langer R, Davies MC, Alexander MR (2008) TOF-SIMS analysis of a 576 micropatterned copolymer array to reveal surface moieties that control wettability. Anal Chem 80:135–142. doi:10.1021/ac071560k

80. Taylor M, Urquhart AJ, Zelzer M, Davies MC, Alexander MR (2007) Picoliter water contact angle measurement on polymers. Langmuir 23:6875–6878. doi:10.1021/la070100j

81. Hook AL, Yang J, Chen X, Roberts CJ, Mei Y, Anderson DG, Langer R, Alexander MR, Davies MC (2011) Polymers with hydro-responsive topography identified using high throughput AFM of an acrylate microarray. Soft Matter 7:7194–7197. doi:10.1039/c1sm06063e

82. Tweedie CA, Anderson DG, Langer R, Van Vliet KJ (2005) Combinatorial material mechanics: high-throughput polymer synthesis and nanomechanical screening. Adv Mater 17:2599–2604. doi:10.1002/adma.200501142

83. Taylor M, Urquhart AJ, Anderson DG, Langer R, Davies MC, Alexander MR (2009) Partial least squares regression as a powerful tool for investigating large combinatorial polymer libraries. Surf Interface Anal 41:127–135. doi:10.1002/sia.2969

84. Hook AL, Scurr DJ, Burley JC, Langer R, Anderson DG, Davies MC, Alexander MR (2013) Analysis and prediction of defects in UV photo-initiated polymer microarrays. J Mater Chem B 1:1035–1043. doi:10.1039/c2tb00379a

85. Epa VC, Yang J, Mei Y, Hook AL, Langer R, Anderson DG, Davies MC, Alexander MR, Winkler DA (2012) Modelling human embryoid body cell adhesion to a combinatorial library of polymer surfaces. J Mater Chem 22:20902–20906. doi:10.1039/c2jm34782b

Chapter 5
Single-Cell Microfluidic Cytometry for Next-Generation High-Throughput Biology and Drug Discovery

Westbrook M. Weaver and Robert Damoiseaux

Abstract The cell is the smallest unit of life. Commensurate with its importance in biology and medicine, it has traditionally been the focus of technologies seeking to add to our understanding of physiological processes relating to life, death, and disease. Over the past decade, our understanding of cellular complexity has been bolstered by the advent of increasingly precise techniques for the investigation of cellular phenomena. Microfluidic cell cytometry combines analysis on the single cell level with integrated separation and processing techniques and has emerged as one of the most powerful techniques in this context. This chapter will encompass an overview of the current landscape and novel trends as well as challenges facing the technologies in this new field.

Abbreviations

CCD   Charge-coupled device
CD   Cluster of differentiation
CDR   Complementarity determining region
CMOS   Complementary metal-oxide semiconductor
CTC   Circulating tumor cell
DC   Deformability cytometry
DEAL   DNA-encoded antibody libraries

W.M. Weaver
Department of Bioengineering, University of California, Los Angeles, CA, USA

R. Damoiseaux (✉)
Molecular Screening Shared Resources, California NanoSystems Institute,
570 Westwood Plaza, Room 2145, UCLA Receiving, Los Angeles, CA 90095, USA
e-mail: rdamoiseaux@mednet.ucla.edu

E. Palmer, *Cell-Based Microarrays: Review of Applications,*
Developments and Technological Advances, SpringerBriefs in Cell Biology,
DOI 10.1007/978-1-4939-0594-2_5, © Ella Palmer 2014

DNA	Deoxyribonucleic acid
ELISA	Enzyme-linked immunosorbent assay
EMT	Epithelial to mesenchymal transition
EpCAM	Epithelial cell adhesion molecule
FOV	Field of view
HAI	Hospital-acquired infection (also healthcare-associated infection)
HCS	High-content screening
HIV	Human immunodeficiency virus
HTS	High-throughput screening
IFNγ	Interferon gamma
ISAAC	Immuno-spot array assay chip
MCF7	Michigan Cancer Foundation-7
MET	Mesenchymal to epithelial transition
MITOMI	Mechanically induced trapping of molecular interaction
PCR	Polymerase chain reaction
PMA	Phorbol 12-myristate 13-acetate
POC	Point-of-care
RNA	Ribonucleic acid
RT-PCR	Reverse transcription polymerase chain reaction
SCBC	Single-cell barcode chip
Simoa	Single molecule array
STEAM	Serial time-encoded amplified microscopy
TF	Transcription factor
TNFα	Tumor necrosis factor alpha

5.1 Introduction

Undoubtedly, as the length of scale of interrogation decreases from tissues to single cells, we find that the complexity of multicellular entities is much greater than the sum of their parts one would have assumed. One of the main problems in translational research is the fact that an aggregate measurement of cells (e.g., in a microtiter plate) does not reflect the uniqueness of cell subpopulations, let alone single cells, which might have unique or rare phenotypes. However, these rare cells can often play a crucial role in biology and are quite different from overall phenotypes present in a population of cells. Furthermore, cells behave very differently when they are embedded in an interactive structure such as tissue, where cells are stimulated by their microenvironment and in turn also contribute to the mechanobiological and overall state of the microenvironment. Environmental cues such as tissue mechanics, soluble factors, and fluidic forces have been implicated in the initiation and/or progression of normal physiology such as tissue development [1–3] as well as a number of disease states, e.g., tumorigenesis [4,5] or even instances of bacterial pathogenesis [6].

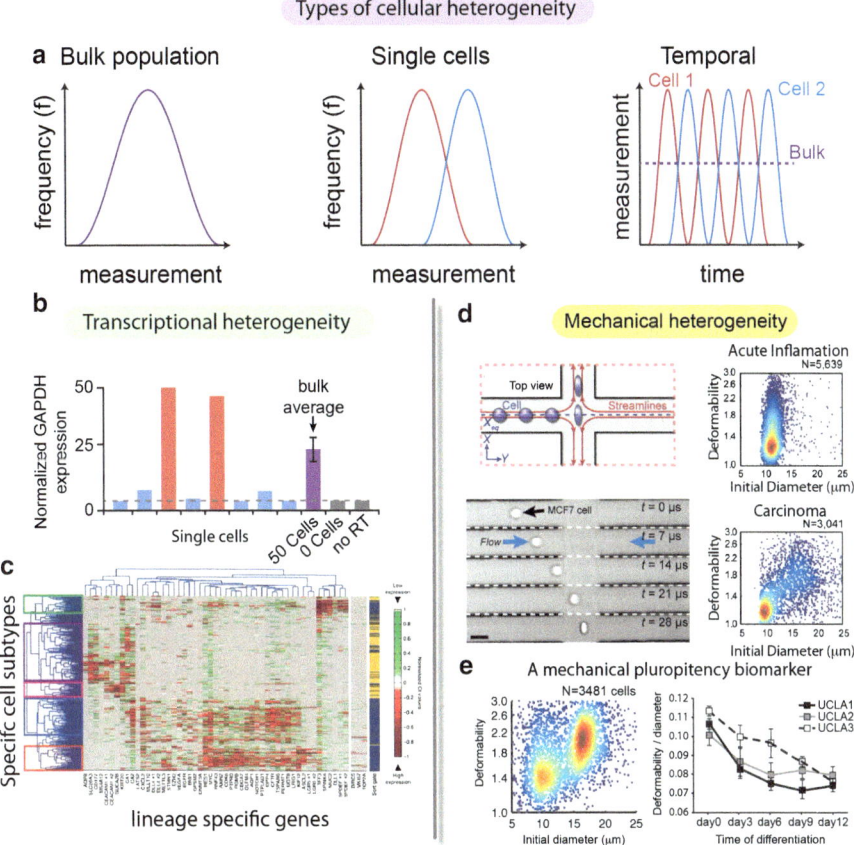

Fig. 5.1 (**a**) Potential ways in which cellular heterogeneity can be manifested in the measurement of a bulk cell population. (**b**) Single cell RT-PCR using microfluidics has shown directly that previously described "incomplete" siRNA knockdown is the result of pooling the heterogeneous response of a cell population in which some cells display complete knockdown, whereas others display only ~50 %. (**c**) Transcriptional heterogeneity within a colon tumor biopsy at the single cell level. Delineations between subtypes are illustrated in this heat map representation. (**d**) Mechanical biomarkers measured in high throughput are indicative of cell subtype. Measurement of pleural fluid effusions from human patients delineates between inflammation and carcinoma prognoses. (**e**) Pluripotency is also linked to cellular deformability, and quantitation of deformability can predict differentiation state

5.1.1   Single Cells Drive Population Phenotypes

The influence of single cells on a large population can be seen in many instances of biology (Fig. 5.1). For example, a normal functional human immune system contains, at any given moment, millions of B-cells, producing discrete subsets of antibodies that comprise human-acquired immune recognition [7]. Further, upon infection, T-cells secrete a suite of interleukins and cytokines in a polymorphic and dynamic fashion, where each step is required for a proper immune response. Transcriptional profiles of bulk cells have been found to be the sum of digitized "on/off" states of individual cells (Fig. 5.1b, c). Even bacterial cells, typically thought of as purely single-celled organisms, often come together in communities of multiple species and act as a pseudo tissue [8,9], secreting extracellular matrix and modulating metabolic profiles to increase fitness and survival [10,11]. This process requires seemingly clonal populations to develop into distinct heterogeneous phenotypic subsets of cells.

Therefore, an understanding of the heterogeneity of a cell population is paramount for our full understanding of any given biological process. Moreover, resolution down to the single cell level is important for diagnostic purposes and also to increase our understanding of the underlying disease to better inform treatments and identify therapeutic targets. In this vein, the development of high throughput, quantitative methods of investigation on the single cell level is a necessity, due both in part to the need for identification of small subsets of cell populations, and biological variation between patient samples frequently encountered in the clinic. At the same time, the rareness of a cell population can make it difficult to obtain statistically significant results and gain insight into, e.g., regulatory pathways involved in physiology. High throughput microfluidic cell cytometry introduces a new paradigm of measurement, moving from standard well plate formats toward serial flow-through processing as well as miniaturization and parallelization of the well plate concept on chip (Fig. 5.2).

5.1.2   Microfluidics Break the Fundamental Limitations of HTS

Of course, microtiter plate screening methods and conventional high throughput and high content screening (HCS) methods have been very useful for drug discovery and basic research over the past few decades, leading to numerous improvements in medicine and significant additions to our biological understanding. These techniques are useful in this context because of their robust and simple nature, as well as the fact that the high throughput screening (HTS) community has evolved around the microtiter plate, resulting in technologies tailored to their shape, size, and functionality. The limitations of the microtiter plate format become evident when we consider the detection of rare subpopulations and rare cellular events: in biological measurement, statistics are extremely important due to biological variation, stemming both from the inability of these experiments to control for all variables and the heterogeneity of cell populations due to cell cycle state, stem cell populations, and other distinct subpopulations even within clonal cell samples. Simple calculations regarding the frequency of a rare cell (perhaps 1 % of the whole population: a generous number) in a confluent 384 well plate (about 5,000 mammalian cells total) lead us to the realization that, in each well, we could only hope to capture the phenotype of 50 cells in a best-case scenario. In HTS applications where hundreds of thousands

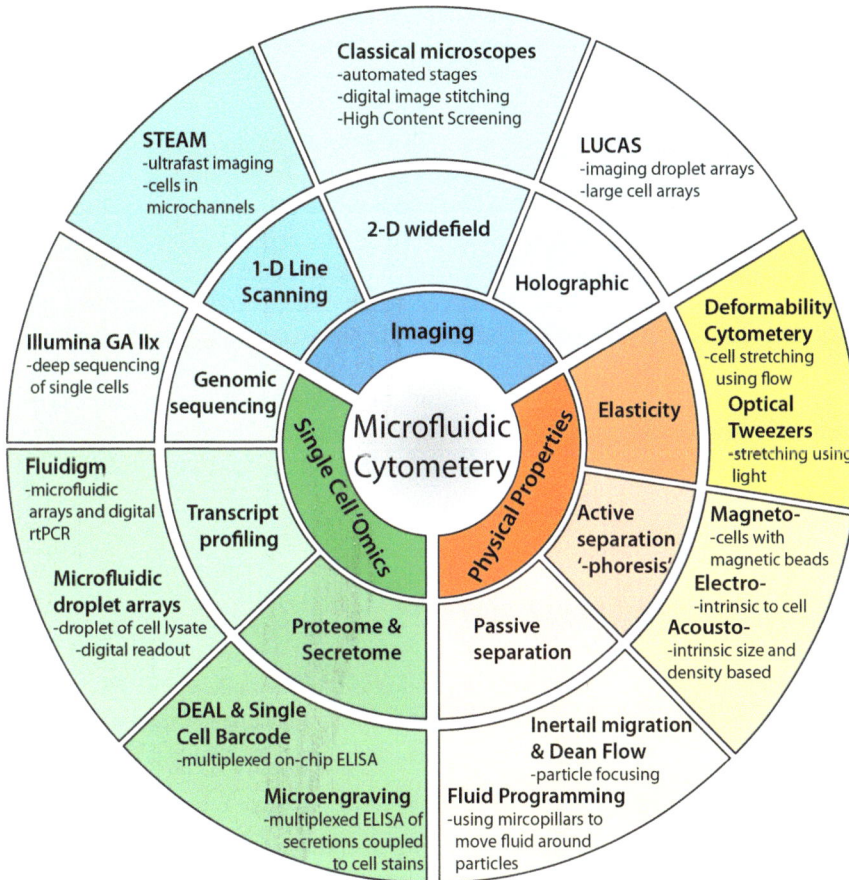

Fig. 5.2 The components of microfluidic cytometry. Technical components such as microfluidic liquid handling and measurement, coupled with novel-imaging techniques are enabling this emerging field. Significant advancement has been made in the fields of proteome profiling of individual cells isolated from carcinomas, secretome profiling of immune cells, transcriptional and genomic profiling of individual cells, as well as the development of mechanical biomarkers for cell state and function

of compounds are to be screened, an enlargement of sample size is not practical and we are left with a dangerously low sample size for reliable measurement. This issue can be remedied to some extent by HCS approaches (allowing multiparameters analysis) and replicate wells (increased sample size), but the fundamental statistical problem of small sample size persists in HTS and also diagnostics, where cell populations of interest might be even much rarer than 1 % of the total cell. Two possible solutions to this problem are either a switch to serial processing of a large number of single cells, or combining enrichment of rare cells with miniaturization of surface assays using, e.g., cell-based microarrays. Microfluidic cell cytometry platforms offer potential methodologies for both of these approaches.

In fact, several powerful methods of single cell analysis have emerged over the past few years, building on classical methods of molecular and cellular analysis techniques. These microfluidic cytometry technologies can be organized based on the measurement type of (1) classical fluorescent and brightfield imaging, (2) physical characterization of single cells, and (3) single cell genomic transcriptomic and proteomic profiling, as depicted in Fig. 5.2. Single cell deep sequencing and reverse transcription-PCR (RT-PCR), whole blood and single cell proteome profiling, secretome arrays, and microfluidic high-speed serial processing and separation of cell types have brought the scientific community a deeper understanding of the importance of cell heterogeneity. New biomarkers based on cellular mechanics have been identified as potential prognostic and diagnostic measurements. In this chapter we will focus primarily on these recent technologies in the context of their biological question, and how they have enabled potential diagnostic tools and furthered scientific understanding of complex biology. We will conclude this chapter with a closer look at platform technologies, translational applications, and future trends of this exiting technology.

5.2 Tumorigenesis and Cancer Biology

Cancer itself cannot be defined as a single disease, but is rather an umbrella term, describing an ever-increasing set of characteristics and cellular aberrations that ultimately lead to tissue neoplasia, metaplasia, and tumor formation [12]. It is this tumor formation that results in dysregulation of the normal tissue equilibrium and eventually organ shutdown.

Differences aside, there is an underlying theme between many cancer types, which is the existence of a stem-cell like subpopulation [13,14], giving rise to new progeny and may lead to the neoplastic behavior, as well as the circulating tumor cell (CTC), ultimately responsible for tumor metastasis by detaching from the original tumor, travelling through the vasculature and exiting into the interstitium at a new site. The stem-cell like population is thought to be formed by dedifferentiation of existing tumor cells as well as the activation of epithelial–mesenchymal transition (EMT), and the reverse mesenchymal–epithelial (MET) transformation [15]. These processes are not yet fully understood, but the current paradigm suggests that EMT occurs both in the tumor mass and during the processes of intravasation into the blood stream [16], most likely leading to a dedifferentiated state similar to a Twist 1-mediated stemness [17]. Conversely, MET, a redifferentiation process, is thought to occur downstream during extravasation to a metastatic site and may be required for cellular proliferation and metastatic formation [18].

Both cancer stem cells and CTCs play a pivotal role in disease progression. Cancer stem cells (perhaps having undergone an EMT) residing in the tumor mass may contribute significantly to drug resistance and relapse after treatment regimes [19,20], and CTCs are implicated in metastasis and hence disease progression [21]. Both of these cell types are small fractions of the cancer cell population. In the case of the CTC, once in the blood, it is an extremely rare cell—sometimes as low as 1 cell in 1 mL of patient blood containing approximately 10^9 normal red blood cells and white blood cells.

5.2.1 Single-Cell Genotyping and Mechanophenotyping

Microfluidic cell cytometry-based detection and profiling of cell types from the primary tumor mass, as well as the CTCs in the bloodstream, offers a very powerful tool for monitoring the progression event as well as dissecting the mechanisms of progression itself. For example, our understanding of the tumorigenic process is enhanced by single cell transcriptional analysis. This microfluidic cytometry technique has shown that colon tumors exhibit the transcriptional diversity of different cellular lineages in a normal colon (Fig. 5.1c). Further, monoclonal tumors from a single ($n = 1$) cell xenograft, still exhibit this transcriptional pattern of multiple cellular lineages, giving strong evidence for the hypothesis of a stem-like population within the tumor mass and retention of this property within this cell subpopulation [22]. This single cell transcription profiling technique utilizes integrated microfluidic circuits of pressurized valves and individually addressable chambers.

When moving away from the primary tumor mass, cellular populations residing in blood or pleural cavity effusions have to be considered. In this context, high throughput microfluidic flow-through techniques have proven to be very useful. Among these techniques is deformability cytometry, a technique in which whole cells are serially deformed in an extensional flow region at a rate of 1,000s per second (Fig. 5.1d) [23]. This method has indicated that deformability may be an effective biomarker for identification of acute immune responses, as well as indication of prognosis for varying cancer types. Complementation of conventional cytological analysis of cell smears and pathology performed on tissue blocks with a quantitative method such as deformability cytometry may help significantly reduce the instance of "atypical" cytological cases in the clinic. These "atypical" cases often result from the lack of clear indicators of malignancy such as large nuclear to cytoplasmic ratio and abnormal nuclear morphology [24]. These cases are notoriously hard to diagnose in early stages, often resulting in tumor progression before confirmed diagnosis, leading to a poor prognosis before treatment can even begin. In fact, deformability cytometry uses mechanophenotyping of pleural cavity effusions to quantitatively diagnose these "atypical" cases as either malignant or benign prior to a conventional cytological follow-up visit, which involves expensive molecular tests and secondary tissue biopsies [25].

5.2.2 Capture and Analysis of CTCs from Blood

Moreover, the isolation of rare CTCs has been accomplished using flow-through microfluidics. The methodologies employed include surface marker-based methods utilizing separation based on magnetic beads [26–28] or posts [29]. These shaped objects can be coated with capture antibodies against cells of epithelial origin, which are not normally found in blood. Another approach utilizes nanomaterial-based capture that exploits the aberrant cell surface structure of a CTC. Here, the cell surface interacts directly with the nanomaterial in a Velcro-like fashion [30]. Magnetic methods are specific; however, somewhat lower throughput (15 min for 200 μL of blood) than other methodologies as high flow rates impart fluidic forces on the cells that can overcome the magnetic forces holding the magnetic beads in place. Posts coated with antibodies have proven to be effective for capturing cells; however, the subsequent release has proven nontrivial. A high flow rate CTC capture technique by micro-vortices in sudden expansion and contraction chambers has shown promise for high throughput CTC isolation, using inertial forces for size-based separation in flow [31,32]. This method also allows easy release of concentrated CTCs in high purity (>90 % purity) for downstream analysis [33]. The ability of captured cells to be easily released for downstream processing will enable both scientific discovery and clinical diagnosis. Combination of these capture techniques with single cell sequencing, RT-PCR and immunocytological staining on chip will enable extremely accurate determination of CTC state, and perhaps (when combined with knowledge of EMT and MET and their effect on tumor progression) will enable better diagnosis and more effective treatment regimes.

While not strictly cytometry based, we would like to mention as well other detection methodologies which we expect to have an impact on this field in the near future. Proteomic profiling of both whole blood and, more recently, single cells has shed tremendous light on the cellular states that define certain subpopulations of cells in tumorigenesis. The single cell barcode chip (SCBC) [34] utilizes DNA-encoded antibody library (DEAL) [35] arrays, combined with sandwich ELISA on the channel surface for multiplexed detection of up to 12 plasma proteins for whole blood, or cellular proteins (both cytosolic and membrane bound) in the case of the single (up to 5) cell chip. The SCBC has enabled a more comprehensive understanding of altered signaling states under treatment with erlotinib and emphasized heterogeneity with respect to this response between single glioblastoma cells. This insight could enable a better understanding of drug treatment options specific to cell state, novel targets, and pathways of interest for the cancer cell. Moreover, these data could be used to better direct the usage of combination therapy such as drug cocktails, even rational design, or at least selection of future drugs for use in the patient.

5.3 Bacterial and Viral Pathogenesis

Bacterial pathogenesis is a complex problem to approach due to the wide variety of important infectious organisms, and the myriad of mechanisms utilized by various bacterial species for infectivity. The opportunistic microbes responsible for hospital-acquired infections (HAIs) resulting from implanted medical devices are of particular mention in part because of the significant impact they have on the healthcare system today [36,37] and the increasing difficulty in treatment because of antibiotic resistance [38,39].

5.3.1   Nanoliter Encapsulation of Bacterial Cells

The question of how antibiotic resistance arises in a population of bacteria has been recently addressed utilizing microfluidic two-phase plug-based systems. These are flow-through, serial systems, where large volumes (up to liters) of bacterial solutions, or even raw ecological samples such as pond water, are in turn compartmentalized into nanoliter droplets in microfluidic channels [40]. The tenability of these systems has been further increased by the introduction of secondary and tertiary solutions to the plug flows by mixing of the miscible stock aqueous solutions from multiple fluidic inlets into combined plugs-in-oil flows (see Fig. 5.3), allowing bacteria to be compartmentalized with varying concentrations and types of antibiotics, as well as the readout for measurement: typically a DNA-intercalating dye (Fig. 5.3). Subsequently, downstream these plugs are quantified using fluorescence microscopy [41]. These systems have directly shown evidence for the founder effect, in which a single bacterium from a clonal population acquires resistance to an antibiotic and creates a new clonally distinct population (Fig. 5.3b) [42].

The ability of certain bacteria to form, and exist in, both mono and multispecies biofilms have long been considered a virulence determinant, especially when considering implantable device-related infections [37]. Further, quorum sensing, the ability of bacteria to "sense" the local density of their inter and intraspecies partners, is thought to control the biofilm forming capability in many of these organisms by genetic regulation which often includes toxin production [43,44]. Microfluidic systems have been well suited for investigation of these phenomena as next-generation high throughput culture systems, incorporate fluid flow with well-defined shear, and spatially organized compartmentalization of bacteria. Using high throughput measurements on microwell arrays confining bacteria spatially while maintaining paracrine-signaling capability, "quorum acting" has been utilized to directly observe such phenomena in B. cereus. Here, high local density (i.e., small distances between microwells) induced transcriptional changes leading to increased clotting capabilities when exposed to whole blood [45].

5.3.2   Viral Load Detection in Fluids

Detecting the presence and amount of viral particles in both patient blood and ecological samples will be paramount for effective diagnosis and monitoring of antiviral treatments. Classically, the enzyme-linked immune-sorbent assay (ELISA),

Fig. 5.3 (a) Stochastic confinement coupled with plug-based microfluidic handling allows for the interrogation of single bacterial responses to antibiotic treatments. (b) Direct observation of the founder effect is seen when bacteria are individually compartmentalized with antibiotic before coming in contact with the drug. (c, d) This system can be used to generate (c) antibiotic resistance and (d) minimum inhibitory concentration profiles for any bacterium, in this case MRSA

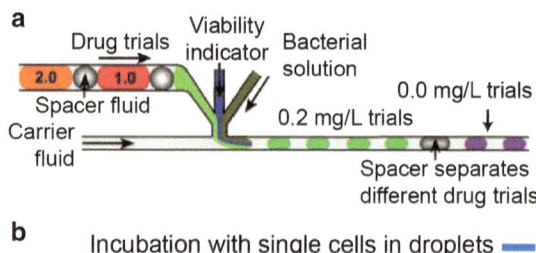

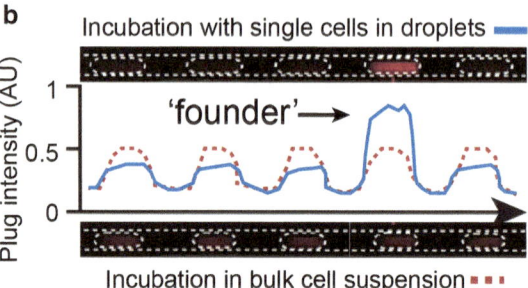

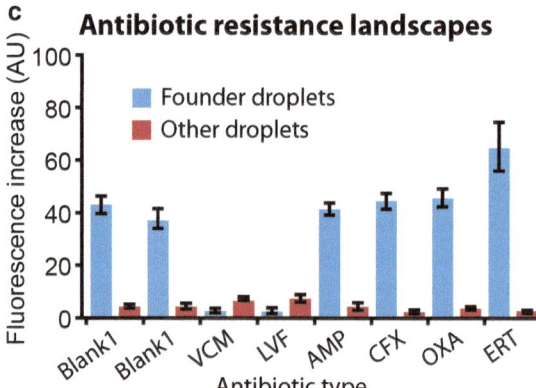

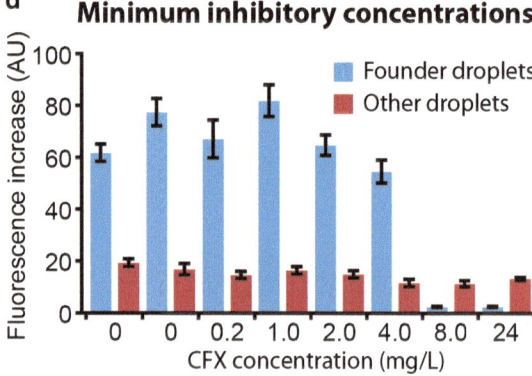

combined with novel PCR methods such as the T7 polymerase technique, rolling circle amplification, and isothermal PCR, have been indispensable tools in detecting, with high sensitivity and specificity, the presence of proteins in solution. The recent advances in high throughput droplet microfluidics and microwell arrays have enabled an even lower limit of detection (approaching the sub-femtomolar concentration regime) while simplifying the readout, through the use of digital ELISA assays [46]. The readout concept is similar to that of digitized PCR and RT-PCR, relying on extremely low dilutions of target molecules, and use of Poisson distributions and statistics. Typically these platforms use beads as the solid phase, where these are then compartmentalized in wells or droplets [47–49]. Further, some point-of-care (POC) devices have been developed using these "ELISA on chip" methods. Specifically, the mChip combines the ease of fluid handing in a lab-on-chip context, with gold nanoparticle-linked ELISA [50]. The readout for this platform is quite clever, in that the assay relies on the growth of silver layers on the gold nanoparticles. If there are gold nanoparticles on the surface (i.e., antigen present), then the absorbance of the silver is measured using inexpensive light-emitting diodes and detectors. This is an excellent example of the miniaturization of a sensitive technique for an affordable and effective POC platform.

5.4 Single-Cell Immunotyping

The human immune system is immensely complex, composed of multiple distinct cell types, residing in different tissues. B cells producing antibodies are responsible for the hosts-acquired immunity and are the mechanisms by which we can become vaccinated against infection from ecological pathogens. Estimations are that at any given point, there are 10^{10}–10^{11} B cells producing at least ~10^{10} distinct antibodies, as measured by deep sequencing of their complementarity determining region $(CDR)_3$ [51]. Further, the pattern of communication between cell types through the secretion of interleukins and cytokines forms a complex network, much of which can be monitored through analysis of blood with the right toolset.

5.4.1 Nanowells and Microengraving for Secretome Analysis

For example, functional cellular immunotyping is a potentially powerful tool for patient monitoring and scientific discovery, enabled primarily by a handful of technologies employing massively parallel arrays of nanowells etched in silicon termed "microengraving" (Fig. 5.4c, d) [52]. This technique combines both the microwell technology with ELISA-like surfaces upon which the wells are "stamped," leaving the imprint of the secretome from the cell in that well. These experiments are typically carried out in two parts: First, cells are seeded on surfaces with

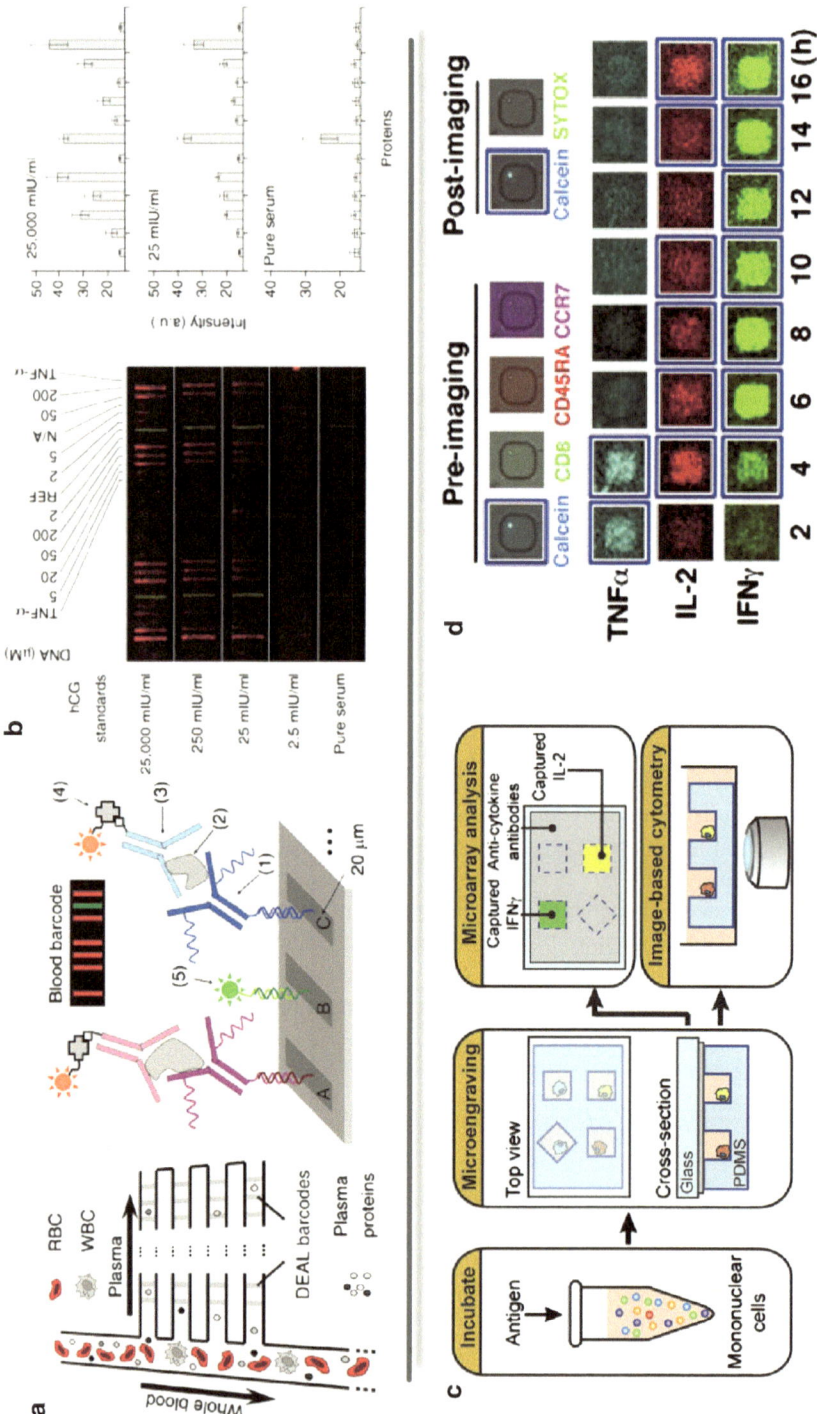

Fig. 5.4 (**a**) Proteomic profiling of whole blood plasma or single cells has been accomplished using a combination of DNA-encoded antibody libraries (DEAL) and valving microfluidics. (**b**) Readout is similar to a barcode, using ELISA to detect captured target protein patterns on the surface. Up to 12 proteins can be profiled simultaneously. (**c**) Secretome profiling of immune cells using a technique termed microengraving has shed tremendous light on the dynamic responses of single T and B cells to stimulation. (**d**) Temporal responses can be easily measured using this technique, coupled with cell staining of identifying CD markers and live/dead staining

highly parallelized etched single cell wells using passive gravity sedimentation. After cell seeding, the substrates can be gently washed and aspirated without removing the cells from their respective well. Subsequently, the cell type of interest (typically T cells) are stimulated by the addition of a small molecule, washed and a glass slide functionalized with antibodies against known secreted factors such as TNFα, various interleukins, and interferons are placed face down over the wells. After a given incubation time of interest, the slide is removed and the cells are fixed and stained for various surface markers such as CD4, CD8, and can even be used with pre-stained cell barcodes for multiplexing [53]. Simultaneously, the cover slide is used as an ELISA substrate to identify components of the secretome of each single cell well where the number of individual components that can be assayed in a single experiment is limited only by the existence of an antibody directed against it, the number of mutually exclusive fluorescent readouts on the imaging system being used, and of course the physical space allowed for the microengraved array. This tool has been extremely powerful in immunophenotyping, as well as uncovering a never before seen phenomenon of the polyfunctional dynamic T cell response to infection challenge: In this scenario, T cells respond asynchronously to activation ex vivo with phorbol 12-myristate 13-acetate (PMA) and ionomycin. Their production of IL-2, TNFα, and IFN-γ, is sequential, and the program repeatable as well as dependent upon differentiation state [54]. Microengraving has also been used to quantitatively characterize the dynamic recovery of HIV-specific CD8+ T cells [55].

Secretions from immune cells have been characterized also by other ELISA-based nanowell technologies. Notably, the immunospot array assay on a chip (ISAAC) utilizes a nanowell array in which the top surface (from which the wells are etched) is coated with antibodies against IFNγ and TNFα [56]. This allows for detection of CD4+ T cells (via cell surface staining) multiplexed with assessment of secretion of one or more cytokines. This technique measures similar analytes/phenotypes as the previously mentioned technique but requires only one substrate. However, the resulting patterns formed on the chips require significant image analysis in order to determine the original well from which the analyte was secreted, as ELISA signals from neighboring wells can overlap (Fig. 5.4d).

5.4.2   Real-Time Detection of Transcriptional Changes in Immune Cells

Other techniques utilizing integrated microfluidic culture systems coupled with fluorescence microscopy and RT-PCR on chip have shed light on how information processing occurs in response to TNF-α [57]. This technique illustrates that NF-kappaB is digitally activated on a cell-by-cell basis in response to a homogenous TNF-α dosage. This observation was previously unobtainable without the use of a high throughput, quantitative single cell analysis using these microfluidic systems.

5.5 Platform Functionalities and Components

Sample image capture and subsequent analysis are classic bottlenecks in HTS system flows, and novel imaging modalities enable many of these next-generation screening platforms to be practically implemented. Imaging techniques can be grossly categorized into three subsets: (1) classical lens-based imaging coupled with automated stages and capture, (2) wide field of view (FOV) holographic techniques, and (3) superfast serial methods. Typical lens-based systems are widely used in conjunction with commercial automated xyz stages for printed microarrays, as well as HCS of both 384 and 1,536 well plate formats. These techniques are well described and will not be the focus here.

5.5.1   Novel Imaging Modalities

Novel lens-free, wide FOV systems make use of a CMOS or CCD pixel array, resulting in overall small instrument size while maintaining large fields of view, up to 10 mm × 10 mm [58,59]. This is a very powerful technique as it allows for capture of all z planes in the FOV because there are no lenses, only direct collection of transmitted light from the sample on the CMOS array. The absence of lenses both enables the wide FOV and lowers the cost. However heavy image postprocessing and reconstruction from the "holographic" image collected is required. Typically, these systems have found use in cell imaging for using brightfield and darkfield imaging especially for counting purposes as well as cell motility and morphology measurements [60,61]. There have been recent efforts toward the development of fluorescent-imaging modalities using lens-free systems, where success has been made and demonstrated with capabilities for resolving single *C. elegans* [62]. These systems will be very well suited for imaging of digitized ELISA and PCR arrays both in droplets and discrete wells, especially since reconstruction is not necessary for data acquisition in these cases. Further, these imagers are easily coupled with microfluidic systems because the chips can simply rest directly on the CMOS array, making integration with droplet generation technologies very simple [63].

Ultrafast serial imaging is utilized in a number of microfluidic platforms relying on fast serial processing of single cells in flow. Commercial high-speed CCD cameras are utilized for a number of applications, including deformability cytometry and most inertial microfluidic techniques; however, these produce a significant cost burden on the researcher or end user. Serial time-encoded amplified microscopy (STEAM) is a novel imaging method using ultrafast line scanning to detect particles and cells in flow [64]. This technique has been used to detect specific cell types coated with beads containing anti-EpCAM antibodies, as well as to identify morphological parameters unique to MCF7 breast cancer cells as compared to normal white blood cells in flow [65]. Coupled with simple straight microfluidic channels designed for particle focusing, this technique has unparalleled throughput, as the speed of the imaging is in the GHz range, allowing throughputs approaching 100,000 cells/s, and extremely low false positive rates of one in a million.

5.5.2   *Microfluidic Interrogation and Separation Techniques*

Microfluidic technology enables such a wide range of novel processes from liquid handling and sample processing to the ability to address single cells in high throughput that it deserves special mention. These technologies utilize a range of force fields to manipulate and address single cells in flow. Magnetic, electric, and even optical fields can be used to manipulate single cells in three dimensions. Magnetic forces are typically used in conjunction with pre-labeling of cells with magnetic particles coated with antibodies directed against a surface marker of interest, followed by passing the cell-bead conjugates through microchannels with embedded magnetic elements [28]. The fields augment the trajectories of cells dependent upon labeling, and then these cells can be siphoned off from channel outlets. Electroosmosis works in a similar fashion, however pre-labeling of cells is not necessary, as the electromobility of cells is an intrinsic property, shown to be dependent upon membrane composition and cell size, thus allowing discrimination between cell types [66,67]. Optical forces have also been employed as a single cell interrogation technique, termed "optical tweezers" [68]. This method traps a cell in a focused optical field, relying on the difference in refractive index between the cell and its surrounding solution. The field can then be tuned to physically stretch the cell, measuring its strain response to a stress field, inferring mechanical properties of the cells such as elastic modulus. These platforms are very precise using highly tunable field forces and gradients; however, their throughput is somewhat limited to about 10 cells per minute at maximum, although a number of massively parallelized optical trap methods are being developed [69].

Converse to the addition of external force fields in microfluidics, the use of the flow fields generated simply by fluid moving through the channels is also a highly effective method for ordering and addressing cells serially in microchannels. Inertial focusing of particles and cells to specific locations within a channel cross-section is a novel technique being adopted by an increasing number of research groups [70]. Briefly, the balance of the shear gradient lift force (stemming from the parabolic velocity profile in a channel) and a wall effect force (possibly from the reflection of viscous wakes generated by the particles themselves) [71] ultimately leads to deterministic positioning of objects in the channel cross-section. Further, this location is dependent on the size, shape, and deformability of the particles in flow. This phenomenon requires the presence of non-negligible inertial forces, which was initially thought to be unobtainable in microflows simply due to the velocity of flow required to reach Reynolds numbers greater than 1 (less than 1 is referred to as Stokes flow, where only viscous forces are considered to be of consequence). Practically speaking, Reynolds numbers of 10 or greater are the operating regimes for inertial microfluidics. This technique offers a very powerful method of passive ordering for imaging and/or separation of particles in flow. One type of flow field, the extensional flow, has been used in conjunction with upstream inertial focusing of cells to the channel center, to deliver cells to a region of extremely high fluid shear, imparting large (~10–100 nN) forces on cells. Deformability cytometry is a microfluidic technique utilizing this very force field, coupled with high-speed brightfield imaging to measure cell deformation [23]. This technique can process cells on the order of 1,000s per second, and generate quantitative plots of deformability, size, as well as a number of cellular characteristics including membrane ruffling and dynamic responses to stretch.

The use of inertial forces in microchannels can also be employed to move both fluid and particles within the cross-section of a channel. It has recently been shown that the presence of non-moving obstacles in flow (as compared to particles or cells, which are moving in flow) coupled with inertial flow (Re > 10), results in nonreversible fluid parcel migration in a channel cross-section [72,73]. This is extremely useful for separation of particles (or cells) from an incoming stream that also contains analysis reagent (such as an antibody stain or small molecule). This effectively washes the particles or cells in flow without the addition of any external force field, and the operation range is at high flow rate, allowing 1,000s of cells per second processing rate, assuming concentrations of 10,000s of cells per ml in the incoming solution.

Together these microfluidic techniques have the ability to transform HTS platforms. Many of these technologies are still in the development stage, although some are in the process of moving to market. The obvious limitation at hand is the integration of these microscale fluidic platforms with current HTS workflows, involving large robotics, microtiter plate storage of solutions and chemicals, as well as current macroscale liquid handling techniques. The true transformative power of the microscale technologies cannot be realized until (1) a standard protocol for integration of macro and micro is set and (2) these research and development level technologies, mostly utilizing PDMS on glass chips, are moved to materials and platforms that are more amenable to scaled up production, such as injection molding of hard polymers and premade integrated circuits.

5.5.3   Macro- and Micro-Integration

The development of true next generation, functional high throughput and HCS methods will be hinged on the proper integration of macroscale automation tools used in screening today, with the highly precise microscale methodologies for measurement reviewed here. This realm has been mostly untouched thus far, aside from one example of integration that has led to an extremely powerful tool.

Microfluidic mechanically induced trapping of molecular interactions (MITOMI) combines the power of macroscale robotics for printing DNA microarray libraries with the highly controlled binding microenvironments of microfluidic large-scale integration [74,75]. Microfluidic large-scale integration utilizes a two-layer channel design built in elastic PDMS to achieve valving capabilities in microfluidics. In this technique, the top channel (or control channel) resides above the bottom channel, separated by a thin PDMS membrane [76]. When the top channel is pressurized with either air or liquid, the PDMS membrane deforms and closes the bottom channel. This valve can be released simply by relieving the pressure applied to the top layer.

In the MITOMI technique, synthetic DNA library arrays are printed and immobilized on epoxy surfaces, followed by channel alignment and bonding. Using the valving techniques offered with large-scale integration, a solution containing transcription factor (TF) proteins is injected into the channels, hydrating the DNA and allowing for interaction between TF and DNA. After incubation, button valves are actuated, trapping the DNA and TF in their equilibrium ratios on an antibody-coated spot on the channel surface (targeting the TF). Subsequently, channels are washed with clean buffer, removing any molecules that were not trapped under the button valve. Measurement of fluorescence of the bound TF (pre-labeled with Bodipy) and the DNA (pre-labeled with Cy5) allows for direct, quantitative measurement of equilibrium conditions, specifically the ratio of the number of TF molecules and DNA molecules by comparing surface fluorescence against a premade calibration curve. This technology has the potential for direct measurement of the dissociation constant (K_D) of the TF and that sequence of DNA and may be applied to any interaction based on a single binding site model.

This methodology has been automated using robotics and microfluidics which lead to insights of the "binding fingerprints" of transcription factors. Further, this platform has been applied to protein recognition of the secondary structures of RNA molecules [77], and one could easily imagine this techniques applicability in any molecular interaction study. These insights will be very useful both in more quantitative understanding of genetic circuit regulation mechanisms and in rationally designing our own synthetic biology circuits.

5.6 Translational Applications

Although research and development of novel technologies in the academic sector is a major driving force for discovery and understanding, their implementation and scale-up to the industrial sector is what brings the technology to the bedside. In order for these "future" technology platforms to really have an effect on treatment efficacy and patient outcome, a compartmentalized, marketable platform must be realized.

Currently, there are a number of examples of commercialized versions of these technologies, including digital PCR and ELISA methods, as well as microfluidic processing and cell handling. These technologies appear to be successful for two main reasons: (1) they are not over-complicated (although there are high levels of functionality in the systems) leading to robust and reproducible behaviors and (2) they have been converted to a "black box" type system, where there are only two main components that the user has to worry about: a measurement box and a disposable cassette. These two aspects allow for relatively easy integration into current workflows of HTS and HCS systems. Further, many companies have sized their cassettes to match the microtiter plate format, allowing for direct integration with current robotic handling systems.

There are a few companies of note that have successfully commercialized a microfluidic technology, imaging technology, or a combination of the two. Both digital ELISA and digital PCR methods have become readily available to researchers over the past few years. Quanterix's Simoa (single molecule array) technology, utilizing nanowells for molecule capture, sealed with an oil layer flowed over the substrate, has been successful in the areas of viral and bacterial pathogen detection, resulting mainly from the extremely low level of detection the system offers. Commercialized digital PCR methods include Fluidigm [78], using integrated microfluidic valves and digital PCR readout, and Raindance [79] as well as QuantaLife [80] (which was recently acquired by Bio-Rad) both employing droplet-based microfluidic systems for compartmentalization of the sample solutions.

There are also many companies that are still in the start-up stages of development. Companies utilizing microfluidic methods include Cytovale [23], one of the first companies harnessing inertial microfluidics in a commercialized product for cellular deformability evaluation. This company aims to translate the deformability cytometry platform discussed in the tumorigenesis and cancer biology section of this chapter. Imaging-based technologies are also being commercialized, most notably Holombic [58], commercializing the LUCAS holographic imaging system, as well as Nantwork, who has acquired the STEAM [64] ultrafast line-scanning imaging technology. Although both of these companies are commercializing novel imaging modalities, they are pursuing very different markets, where Holombic is geared toward the simple, POC tool for diagnostics in the field and the STEAM technology may be more fitting for a research or core lab setting.

5.7   Conclusion and Future Directions

The burgeoning field of single-cell microfluidic cytometry has not only provided the scientific community with a new tool set for quantitative biological investigation, but has already led to novel insight such as the discovery of rare cell phenotypes previously unresolvable. The ability of these techniques to miniaturize and parallelize well plate techniques using microarrays or serially process large numbers of single cells in high throughput has been an enabler for this push forward in scientific discovery.

A key challenge for the future for these microfluidic-based assays is their integration with existing HTS workflows. Microfluidic cytometry devices can empower HTS and HCS by adding novel readouts including single cell biomechanical measurements, biomarker development and detection as well as single cell genomics and proteomics. These measurements could transform our understanding of the effects of small molecule libraries on cell function, and greatly increase the efficacy of treatments derived from these screens.

It is likely that, in the next decade, both basic science and drug discovery pipelines will see some dramatic changes with functional integration of these novel technologies into the workflows. If microfluidic cell cytometry toolsets are thoughtfully developed, with HTS and HCS integration in mind, then it is reasonable to predict an exponential increase in data throughput and content, creating a near future where understanding the connectivity and dynamics of every component of a system as complex as a multicellular eukaryotic structure is possible.

References

1. Watt FM, Huck WTS (2013) Role of the extracellular matrix in regulating stem cell fate. Nat Rev Mol Cell Biol 14:467–473. doi:10.1038/nrm3620
2. Miller CJ, Davidson LA (2013) The interplay between cell signalling and mechanics in developmental processes. Nat Rev Genet 14:733–744. doi:10.1038/nrg3513
3. Wozniak MA, Chen CS (2009) Mechanotransduction in development: a growing role for contractility. Nat Rev Mol Cell Biol 10:34–43. doi:10.1038/nrm2592
4. Jaalouk DE, Lammerding J (2009) Mechanotransduction gone awry. Nat Rev Mol Cell Biol 10:63–73. doi:10.1038/nrm2597
5. Butcher DT, Alliston T, Weaver VM (2009) A tense situation: forcing tumour progression. Nat Rev Cancer 9:108–122. doi:10.1038/nrc2544
6. Weaver WM, Milisavljevic V, Miller JF, Di Carlo D (2012) Fluid flow induces biofilm formation in Staphylococcus epidermidis polysaccharide intracellular adhesin-positive clinical isolates. Appl Environ Microbiol. Available: http://www.ncbi.nlm.nih.gov/pubmed/22706049. Accessed 10 July 2012.
7. Jiang N, He J, Weinstein JA, Penland L, Sasaki S et al (2013) Lineage structure of the human antibody repertoire in response to influenza vaccination. Sci Transl Med 5:171ra19. doi:10.1126/scitranslmed.3004794
8. McDougald D, Rice SA, Barraud N, Steinberg PD, Kjelleberg S (2012) Should we stay or should we go: mechanisms and ecological consequences for biofilm dispersal. Nat Rev Microbiol 10:39–50. doi:10.1038/nrmicro2695
9. Davies D (2003) Understanding biofilm resistance to antibacterial agents. Nat Rev Drug Discov 2:114–122. doi:10.1038/nrd1008
10. Jones S (2007) Biofilms: How does your biofilm grow? Nat Rev Microbiol 5:168–169. doi:10.1038/nrmicro1626
11. Flemming H-C, Wingender J (2010) The biofilm matrix. Nat Rev Microbiol 8:623–633. doi:10.1038/nrmicro2415
12. Potter JD (2007) Morphogens, morphostats, microarchitecture and malignancy. Nat Rev Cancer 7:464–474. doi:10.1038/nrc2146
13. Feinberg AP, Ohlsson R, Henikoff S (2006) The epigenetic progenitor origin of human cancer. Nat Rev Genet 7:21–33. doi:10.1038/nrg1748
14. Dalerba P, Cho RW, Clarke MF (2007) Cancer stem cells: models and concepts. Annu Rev Med 58:267–284. doi:10.1146/annurev.med.58.062105.204854
15. Brabletz T (2012) EMT and MET in metastasis: where are the cancer stem cells? Cancer Cell 22:699–701. doi:10.1016/j.ccr.2012.11.009
16. Thiery JP, Acloque H, Huang RYJ, Nieto MA (2009) Epithelial-mesenchymal transitions in development and disease. Cell 139:871–890. doi:10.1016/j.cell.2009.11.007
17. Tsai JH, Donaher JL, Murphy DA, Chau S, Yang J (2012) Spatiotemporal regulation of epithelial-mesenchymal transition is essential for squamous cell carcinoma metastasis. Cancer Cell 22:725–736. doi:10.1016/j.ccr.2012.09.022
18. Korpal M, Ell BJ, Buffa FM, Ibrahim T, Blanco MA et al (2011) Direct targeting of Sec23a by miR-200 s influences cancer cell secretome and promotes metastatic colonization. Nat Med 17:1101–1108. doi:10.1038/nm.2401
19. Magee JA, Piskounova E, Morrison SJ (2012) Cancer stem cells: impact, heterogeneity, and uncertainty. Cancer Cell 21:283–296. doi:10.1016/j.ccr.2012.03.003

20. Hölzel M, Bovier A, Tüting T (2013) Plasticity of tumour and immune cells: a source of heterogeneity and a cause for therapy resistance? Nat Rev Cancer 13:365–376. doi:10.1038/nrc3498
21. Labelle M, Begum S, Hynes RO (2011) Direct signaling between platelets and cancer cells induces an epithelial-mesenchymal-like transition and promotes metastasis. Cancer Cell 20:576–590. doi:10.1016/j.ccr.2011.09.009
22. Dalerba P, Kalisky T, Sahoo D, Rajendran PS, Rothenberg ME et al (2011) Single-cell dissection of transcriptional heterogeneity in human colon tumors. Nat Biotechnol 29:1120–1127. doi:10.1038/nbt.2038
23. Gossett DR, Tse HTK, Lee SA, Ying Y, Lindgren AG et al (2012) Hydrodynamic stretching of single cells for large population mechanical phenotyping. Proc Natl Acad Sci U S A 109:7630–7635. doi:10.1073/pnas.1200107109
24. Di Carlo D (2012) A mechanical biomarker of cell state in medicine. J Lab Autom 17:32–42. doi:10.1177/2211068211431630
25. Tse HTK, Gossett DR, Moon YS, Masaeli M, Sohsman M et al (2013) Quantitative diagnosis of malignant pleural effusions by single-cell mechanophenotyping. Sci Transl Med 5:212ra163. doi:10.1126/scitranslmed.3006559
26. Kim S, Han S-I, Park M-J, Jeon C-W, Joo Y-D et al (2013) Circulating tumor cell microseparator based on lateral magnetophoresis and immunomagnetic nanobeads. Anal Chem 85:2779–2786. doi:10.1021/ac303284u
27. Xu W, Cao L, Chen L, Li J, Zhang X-F et al (2011) Isolation of circulating tumor cells in patients with hepatocellular carcinoma using a novel cell separation strategy. Clin Cancer Res 17:3783–3793. doi:10.1158/1078-0432.CCR-10-0498
28. Kang JH, Krause S, Tobin H, Mammoto A, Kanapathipillai M et al (2012) A combined micromagnetic-microfluidic device for rapid capture and culture of rare circulating tumor cells. Lab Chip 12:2175–2181. doi:10.1039/C2LC40072C
29. Nagrath S, Sequist LV, Maheswaran S, Bell DW, Irimia D et al (2007) Isolation of rare circulating tumour cells in cancer patients by microchip technology. Nature 450:1235–1239. doi:10.1038/nature06385
30. Lu Y-T, Zhao L, Shen Q, Garcia MA, Wu D, et al. (n.d.) NanoVelcro Chip for CTC enumeration in prostate cancer patients. Methods. Available: http://www.sciencedirect.com/science/article/pii/S104620231300220X. Accessed 29 October 2013.
31. Hur SC, Mach AJ, Di Carlo D (2011) High-throughput size-based rare cell enrichment using microscale vortices. Biomicrofluidics 5. Available: http://www.ncbi.nlm.nih.gov/pmc/articles/PMC3171489/. Accessed 29 October 2013.
32. Mach AJ, Kim JH, Arshi A, Hur SC, Carlo DD (2011) Automated cellular sample preparation using a Centrifuge-on-a-Chip. Lab Chip 11:2827–2834. doi:10.1039/C1LC20330D
33. Sollier E, Go DE, Che J, Gossett DR, O'Byrne S, et al. (2013) Size-selective collection of circulating tumor cells using Vortex technology. Lab Chip. Available: http://pubs.rsc.org/en/content/articlelanding/2014/lc/c3lc50689d. Accessed 29 October 2013.
34. Shi Q, Qin L, Wei W, Geng F, Fan R et al (2012) Single-cell proteomic chip for profiling intracellular signaling pathways in single tumor cells. Proc Natl Acad Sci U S A 109:419–424. doi:10.1073/pnas.1110865109
35. Fan R, Vermesh O, Srivastava A, Yen BKH, Qin L et al (2008) Integrated barcode chips for rapid, multiplexed analysis of proteins in microliter quantities of blood. Nat Biotech 26:1373–1378. doi:10.1038/nbt.1507
36. Chambers HF, DeLeo FR (2009) Waves of resistance: Staphylococcus aureus in the antibiotic era. Nat Rev Microbiol 7:629–641. doi:10.1038/nrmicro2200
37. Otto M (2009) Staphylococcus epidermidis—the "accidental" pathogen. Nat Rev Microbiol 7:555–567. doi:10.1038/nrmicro2182
38. Dolgin E (2010) Sequencing of superbugs seen as key to combating their spread. Nat Med 16:1054–1054. doi:10.1038/nm1010-1054a
39. Dolgin E (2011) "Game changer" antibiotic and others in works for superbug. Nat Med 17:10–10. doi:10.1038/nm0111-10

40. Chen D, Du W, Liu Y, Liu W, Kuznetsov A et al (2008) The chemistrode: A droplet-based microfluidic device for stimulation and recording with high temporal, spatial, and chemical resolution. Proc Natl Acad Sci 105:16843–16848. doi:10.1073/pnas.0807916105

41. Boedicker JQ, Vincent ME, Ismagilov RF (2009) Microfluidic confinement of single cells of bacteria in small volumes initiates high-density behavior of quorum sensing and growth and reveals its variability. Angew Chem Int Ed Engl 48:5908–5911. doi:10.1002/anie.200901550

42. Boedicker JQ, Li L, Kline TR, Ismagilov RF (2008) Detecting bacteria and determining their susceptibility to antibiotics by stochastic confinement in nanoliter droplets using plug-based microfluidics. Lab Chip 8:1265–1272. doi:10.1039/B804911D

43. Molloy S (2010) Quorum sensing: Setting the threshold. Nat Rev Microbiol 8:388–389. doi:10.1038/nrmicro2374

44. Keller L, Surette MG (2006) Communication in bacteria: an ecological and evolutionary perspective. Nat Rev Microbiol 4:249–258. doi:10.1038/nrmicro1383

45. Kastrup CJ, Boedicker JQ, Pomerantsev AP, Moayeri M, Bian Y et al (2008) Spatial localization of bacteria controls coagulation of human blood by "quorum acting". Nat Chem Biol 4:742–750

46. Rissin DM, Kan CW, Campbell TG, Howes SC, Fournier DR et al (2010) Single-molecule enzyme-linked immunosorbent assay detects serum proteins at subfemtomolar concentrations. Nat Biotechnol 28:595–599. doi:10.1038/nbt.1641

47. Hatch AC, Fisher JS, Tovar AR, Hsieh AT, Lin R et al (2011) 1-Million droplet array with wide-field fluorescence imaging for digital PCR. Lab Chip 11:3838–3845. doi:10.1039/c1lc20561g

48. Teste B, Ali-Cherif A, Viovy JL, Malaquin L (2013) A low cost and high throughput magnetic bead-based immuno-agglutination assay in confined droplets. Lab Chip 13:2344–2349. doi:10.1039/c3lc50353d

49. Sakakihara S, Araki S, Iino R, Noji H (2010) A single-molecule enzymatic assay in a directly accessible femtoliter droplet array. Lab Chip 10:3355–3362. doi:10.1039/C0LC00062K

50. Chin CD, Laksanasopin T, Cheung YK, Steinmiller D, Linder V et al (2011) Microfluidics-based diagnostics of infectious diseases in the developing world. Nat Med 17:1015–1019. doi:10.1038/nm.2408

51. Mathonet P, Ullman CG (2013) The application of next generation sequencing to the understanding of antibody repertoires. Front B Cell Biol 4:265. doi:10.3389/fimmu.2013.00265

52. Love JC, Ronan JL, Grotenbreg GM, van der Veen AG, Ploegh HL (2006) A microengraving method for rapid selection of single cells producing antigen-specific antibodies. Nat Biotechnol 24:703–707. doi:10.1038/nbt1210

53. Yamanaka YJ, Szeto GL, Gierahn TM, Forcier TL, Benedict KF et al (2012) Cellular barcodes for efficiently profiling single-cell secretory responses by microengraving. Anal Chem 84:10531–10536. doi:10.1021/ac302264q

54. Han Q, Bagheri N, Bradshaw EM, Hafler DA, Lauffenburger DA et al (2012) Polyfunctional responses by human T cells result from sequential release of cytokines. Proc Natl Acad Sci 109:1607–1612. doi:10.1073/pnas.1117194109

55. Varadarajan N, Kwon DS, Law KM, Ogunniyi AO, Anahtar MN et al (2012) Rapid, efficient functional characterization and recovery of HIV-specific human CD8+ T cells using microengraving. Proc Natl Acad Sci 109:3885–3890. doi:10.1073/pnas.1111205109

56. Jin A, Ozawa T, Tajiri K, Obata T, Kondo S et al (2009) A rapid and efficient single-cell manipulation method for screening antigen-specific antibody–secreting cells from human peripheral blood. Nat Med 15:1088–1092. doi:10.1038/nm.1966

57. Tay S, Hughey JJ, Lee TK, Lipniacki T, Quake SR et al (2010) Single-cell NF-kappaB dynamics reveal digital activation and analogue information processing. Nature 466:267–271. doi:10.1038/nature09145

58. Ozcan A, Demirci U (2008) Ultra wide-field lens-free monitoring of cells on-chip. Lab Chip 8:98–106. doi:10.1039/b713695a

59. Su T-W, Seo S, Erlinger A, Ozcan A (2008) Multi-angle LUCAS for high-throughput on-chip cytometry. Conf Proc Annu Int Conf IEEE Eng Med Biol Soc IEEE Eng Med Biol Soc Conf 2008:1854–1855. doi:10.1109/IEMBS.2008.4649543

60. Su T-W, Erlinger A, Tseng D, Ozcan A (2010) Compact and light-weight automated semen analysis platform using lensfree on-chip microscopy. Anal Chem 82:8307–8312. doi:10.1021/ac101845q
61. Su T-W, Xue L, Ozcan A (2012) High-throughput lensfree 3D tracking of human sperms reveals rare statistics of helical trajectories. Proc Natl Acad Sci U S A 109:16018–16022. doi:10.1073/pnas.1212506109
62. Coskun AF, Sencan I, Su T-W, Ozcan A (2011) Lensfree fluorescent on-chip imaging of transgenic Caenorhabditis elegans over an ultra-wide field-of-view. PloS One 6:e15955. doi:10.1371/journal.pone.0015955
63. Stybayeva G, Mudanyali O, Seo S, Silangcruz J, Macal M et al (2010) Lensfree holographic imaging of antibody microarrays for high-throughput detection of leukocyte numbers and function. Anal Chem 82:3736–3744. doi:10.1021/ac100142a
64. Goda K, Tsia KK, Jalali B (2009) Serial time-encoded amplified imaging for real-time observation of fast dynamic phenomena. Nature 458:1145–1149. doi:10.1038/nature07980
65. Goda K, Ayazi A, Gossett DR, Sadasivam J, Lonappan CK et al (2012) High-throughput single-microparticle imaging flow analyzer. Proc Natl Acad Sci U S A 109:11630–11635. doi:10.1073/pnas.1204718109
66. Gencoglu A, Olney D, Lalonde A, Koppula KS, Lapizco-Encinas BH (2013) Dynamic microparticle manipulation with an electroosmotic flow gradient with low frequency alternating current dielectrophoresis. Electrophoresis. doi:10.1002/elps.201300385
67. Gossett DR, Weaver WM, Mach AJ, Hur SC, Tse HTK et al (2010) Label-free cell separation and sorting in microfluidic systems. Anal Bioanal Chem 397:3249–3267. doi:10.1007/s00216-010-3721-9
68. Capitanio M, Pavone FS (2013) Interrogating biology with force: single molecule high-resolution measurements with optical tweezers. Biophys J 105:1293–1303. doi:10.1016/j.bpj.2013.08.007
69. Roth KB, Eggleton CD, Neeves KB, Marr DWM (2013) Measuring cell mechanics by optical alignment compression cytometry. Lab Chip 13:1571–1577. doi:10.1039/c3lc41253a
70. Zhou J, Papautsky I (2013) Fundamentals of inertial focusing in microchannels. Lab Chip 13:1121–1132. doi:10.1039/C2LC41248A
71. Lee W, Amini H, Stone HA, Di Carlo D (2010) Dynamic self-assembly and control of microfluidic particle crystals. Proc Natl Acad Sci U S A 107:22413–22418. doi:10.1073/pnas.1010297107
72. Gossett DR, Tse HTK, Dudani JS, Goda K, Woods TA et al (2012) Inertial manipulation and transfer of microparticles across laminar fluid streams. Small Weinh Bergstr Ger 8:2757–2764. doi:10.1002/smll.201200588
73. Amini H, Sollier E, Masaeli M, Xie Y, Ganapathysubramanian B et al (2013) Engineering fluid flow using sequenced microstructures. Nat Commun 4:1826. doi:10.1038/ncomms2841
74. Maerkl SJ, Quake SR (2007) A systems approach to measuring the binding energy landscapes of transcription factors. Science 315:233–237. doi:10.1126/science.1131007
75. Fordyce PM, Gerber D, Tran D, Zheng J, Li H et al (2010) De novo identification and biophysical characterization of transcription-factor binding sites with microfluidic affinity analysis. Nat Biotechnol 28:970–975. doi:10.1038/nbt.1675
76. Hong JW, Studer V, Hang G, Anderson WF, Quake SR (2004) A nanoliter-scale nucleic acid processor with parallel architecture. Nat Biotechnol 22:435–439. doi:10.1038/nbt951
77. Martin L, Meier M, Lyons SM, Sit RV, Marzluff WF et al (2012) Systematic reconstruction of RNA functional motifs with high-throughput microfluidics. Nat Methods 9:1192–1194. doi:10.1038/nmeth.2225
78. Ramakrishnan R, Qin J, Jones RC, Weaver LS (2013) Integrated Fluidic Circuits (IFCs) for digital PCR. Methods Mol Biol Clifton NJ 949:423–431. doi:10.1007/978-1-62703-134-9_27
79. Kiss MM, Ortoleva-Donnelly L, Beer NR, Warner J, Bailey CG et al (2008) High-throughput quantitative polymerase chain reaction in picoliter droplets. Anal Chem 80:8975–8981
80. Gardner SN, Hiddessen AL, Williams PL, Hara C, Wagner MC et al (2009) Multiplex primer prediction software for divergent targets. Nucleic Acids Res 37:6291–6304. doi:10.1093/nar/gkp659

About the Author

Ella Palmer is a molecular biology expert. She undertook her Ph.D. at the University of Cambridge, where she developed a large-scale cell-based microarray and identified potential new genes involved in apoptosis, for which a patent was registered. She has since held postdoctoral positions at Imperial College in London, examining the cell polarity of Drosophila photoreceptors, and at University College London, working on a quantitative systems biology initiative for single cell analysis of the NFAT pathway. Most recently she has been working as an intern at a medical writing company.

E. Palmer, *Cell-Based Microarrays: Review of Applications,*
Developments and Technological Advances, SpringerBriefs in Cell Biology,
DOI 10.1007/978-1-4939-0594-2, © Ella Palmer 2014

Index

E. Palmer, *Cell-Based Microarrays: Review of Applications,*
Developments and Technological Advances, SpringerBriefs in Cell Biology,
DOI 10.1007/978-1-4939-0594-2, © Ella Palmer 2014

99